T0205498

Smart Agriculture

Volume 3

The book series Smart Agriculture presents progress of smart agricultural technologies, which includes, but not limited to, specialty crop harvest robotics, UAV technologies for row crops, innovative IoT applications in plant factories, and big data for optimizing production process. It includes both theoretical study and practical applications, with emphasis on systematic studies. AI technologies in agricultural productions will be emphasized, consisting of innovative algorithms and new application domains. Additionally, new crops are emerging, such as hemp in U.S., and covered as well. This book series would cover regions worldwide, such as U.S., Canada, China, Japan, Korea, and Brazil.

The book series Smart Agriculture aims to provide an academic platform for interdisciplinary researchers to provide their state-of-the-art technologies related to smart agriculture. Researchers of different academic backgrounds are encouraged to contribute to the book, such as agriculture engineers, breeders, horticulturist, agronomist, and plant pathologists. The series would target a very broad audience – all having a professional related to agriculture production. It also could be used as textbooks for graduate students.

Xingye Zhu · Alexander Fordjour · Junping Liu ·
Shouqi Yuan

Dynamic Fluidic Sprinkler and Intelligent Sprinkler Irrigation Technologies

Xingye Zhu
Jiangsu University
Zhenjiang, Jiangsu, China

Junping Liu
Jiangsu University
Zhenjiang, Jiangsu, China

Alexander Fordjour
Civil Engineering Department
Koforidua Technical University
Koforidua, Ghana

Shouqi Yuan
Jiangsu University
Zhenjiang, Jiangsu, China

ISSN 2731-3476 ISSN 2731-3484 (electronic)
Smart Agriculture
ISBN 978-981-19-8321-4 ISBN 978-981-19-8319-1 (eBook)
https://doi.org/10.1007/978-981-19-8319-1

This Springer imprint is published by the registered company Springer Nature Singapore Pte Ltd.
The registered company address is: 152 Beach Road, #21-01/04 Gateway East, Singapore 189721, Singapore

Contents

Chapter 1
Introduction

Abstract The effective use of irrigation water is essential for sustainable agriculture, sprinkler irrigation has high prospects for improving water management in crop production. Sprinkler irrigation can be divided into two categories. It includes rotating sprinkler and fixed sprinkler. The hydraulic performance of sprinklers is usually classified as sprinkler, overlapping uniformity and droplet size distribution.

Keywords Sprinkler irrigation · Hydraulic performance · Nozzle · Water saving

1.1 Research Background

Agriculture is highly dependent on climate in developing countries, changes in climatic variables are high to impact agriculture. A study conducted by [1–3] showed that Changes in climatic variables such as atmospheric carbon dioxide concentration, rainfall, humidity, temperature, etc. are found to distress agricultural output. According to [4–6] report, Projections from climate models suggest that the production of +cereal crops indicate that potential yields are to decrease for the most project due to increase in temperature changes in most tropical and subtropical regions. For example, in Pakistan according to [7–9] there will be a reduction of 50% in crop yield. International Rice Research Institute estimate reduction of 20% in yields, in Central and South Asia, there have been projections indicated that yields might drop down to 30%. According to [10–12] the impact of change in climate will add to the several economic and social challenges already being faced by water management in agricultural areas. Freshwater scarcity is one of the most severe natural resource constraints humanity is facing. Globally agriculture is the major user of fresh water, accounting for about 70% of total withdrawals. Agricultural water usage is essentially driven by irrigation [13–16].

Irrigation, therefore, contributes greatly to food security by stabilizing crop production in dry years, which is particularly important for developing countries not to integrate well into world markets. If water and soil resources are poorly managed, irrigation water may have a negative impact on water availability in ecosystems and other water sectors. Globally, irrigation is by far the largest water sector, accounting for about 90% of the extra evapotranspiration caused by human water use [17–19].

X. Zhu et al., *Dynamic Fluidic Sprinkler and Intelligent Sprinkler Irrigation Technologies*, Smart Agriculture 3, https://doi.org/10.1007/978-981-19-8319-1_1

Water is a renewable resource that cannot be depleted because it is recycled in the global hydrological cycle. Therefore, total water consumption is only a weak indicator of sustainability. More important is where and when water is used. In arid periods, when soil water storage is depleted, or in arid areas where precipitation is generally low, irrigation water demand is higher. Therefore, the highest water demand often occurs at the same time as the water shortage, resulting in a water shortage. The key impacts are the decline of groundwater level, the drying up of rivers, the shrinkage of lakes or the vacancy of reservoirs. If the drought lasts for a long time, the shortage of irrigation water supply will lead to a decrease in crop yield and crop area [20–23]. In semi-arid areas, such as California or Southern Africa, drought events often have a greater impact, but even in more humid areas, such as Western Europe, drought has caused huge economic losses, such as the summer of 2003 and 2018 [24, 25]. Not only has a large amount of water extracted from groundwater or surface water raised concerns about the sustainability of freshwater for irrigation, but also about the huge water supply infrastructure needed for this purpose. It is estimated that 17 million reservoirs with a surface area of more than 500,000 square kilometers have been built for water supply and power generation, many of which are used for irrigation [26–28]. Assuming that the average capacity of the reservoir reaches half of its capacity, the annual evaporation on the surface of these artificial lakes is estimated to be 346 km^3, which exceeds the consumption of water for industry, households, and livestock [12]. Also, dam construction affects most of the big rivers.

In this regard, the effective use of irrigation water is essential for sustainable agriculture. According to [29–31] sprinkler irrigation has high prospects for improving water management in crop production. Sprinkler irrigation can play an important role in the development of irrigation in third world countries if the system is properly selected, designed and operated. With the development of irrigation agriculture in the United States, Europe, Australia and other parts of the world, automatic sprinkler systems have accelerated and revolutionary changes have taken place. Therefore, it is not surprising that the utilization rate of sprinkler irrigation systems has been increasing over the years. For example, in the United States, surface irrigation (gravity methods) decreased from 63% in 1979 to 50% in 1994 [30–32] while drip irrigation (drip irrigation, drip irrigation, and micro-spraying) increased from 0.6% in 1979 to nearly 4% in 1994. The biggest and most obvious change is in the area of sprinkler systems (sprinklers and central fulcrums). The sprinkler system increased from 36% in 1979 to nearly 44% in 1994. According to the annual survey of irrigation journals, from 1985 to 2000, the percentage of sprinkler irrigation areas in the United States increased from 37 to 50% [33, 34]. According to the latest report on irrigation research in the United States, the total irrigation area has increased. The growth of the irrigation area is mainly due to the expansion of sprinkler irrigation and central fulcrum irrigation, and the growth of the micro-irrigation area is small. The surface irrigated area peaked in 1980 but has been declining steadily since then [17]. In Spain, sprinkler irrigation is rapidly replacing surface irrigation systems as a result of the modernization of irrigation schemes. The shift of emphasis from surface irrigation to sprinkler irrigation has led to a reduction in the labour force, an increase in irrigation efficiency and an increase in crop yields. Therefore, in recent years, the types of water sprinklers available in the market have also increased dramatically, ranging from traditional single-nozzle or double-nozzle impact sprinklers (with multiple nozzles) to various types of deflector sprinklers [35–37]. The importance of sprinkler irrigation systems in irrigation has long been recognized in China. In 1954, China first recommended the use of an automatic sprinkler system in Russia [19]. Since then, research institutes have been trying to propose new and improved models of automatic sprinkler systems. Sprinkler hydraulic performance is mainly a function of the sprinkler physical features and the important geometrical parameters will influence hydraulic performances. In most parts of the world, inefficient irrigation methods lead to wastage or excessive use of irrigation water. Less than half of the water on average reaches the crop [1, 6, 11, 12], resulting in reductions in crop yields. It has become important for all factors that influence sprinkler water application efficiency and uniformity in the era of water and energy conservation towards the sustainability of diminishing resources. In recent years, low-pressure water-saving has become an important research content in the field of sprinkler irrigation. To achieve the goal of energy and water-saving, there was a need to come out with a new sprinkler that can operate effectively under low-pressure conditions, since low-pressure sprinkler irrigation is gaining momentum all over of the world. Several theoretical, numerical and experimental studies have been conducted to improve the structural and hydraulic performance of the complete fluidic sprinkle. Gan and Chen [38] invented self—controlled step- by complete fluidic sprinkler in1980. Zhu et al. [39] compared fluidic and impact sprinklers and variations in completion time through the quadrants were higher for the fluidic sprinkle. Similarly, Dwomoh et al. [40] carried out a study on the fluidic sprinkler and confirmed the need to optimize the structure. Zhang et al. [41] analyzed the irrigation uniformity of complete fluidic sprinkler in a no-wind condition. Zhu et al. [42] studied the effect of a complete fluidic sprinkler on the hydraulic characteristics based on some important geometrical parameters. Li et al. [43] studied the hydraulic characteristics on the fluidic element of the fluidic sprinkler controlled by clearance· Liu et al. 2008 carried out an experimental study on the range and spraying uniformity influencing factor of complete fluidic· Li et al. [44] studied the wall-attachment of the fluidic sprinkler. Yuan et al. [45] carried out a study on numerical simulation on a new type variable rate fluidic sprinkler. Previous studies have investigated the wall-attaching offset of a fluidic sprinkler. For example, Li et al. [46] studied the theoretical and experimental study on water offset flow in the fluidic component of fluidic sprinklers Hu et al. [47] carried out a study on the fluidic sprinkler and confirmed the need to optimize the structure. Liu et al. [48] researched a numerical simulation and experimental study on a new type of variable-rate fluidic sprinkler. Xu et al. [49] studied numerical simulation and optimization of structural parameters. Their researcher works revealed that when the primary flow jet becomes reattached to the right side, the pressure in the two sidewalls of the main jet flow exclusively depends on flowing duct length and operating pressure. Song et al. [50] studied the flow and heat transfer characteristics of the 2D jet. Their study demonstrated that the structural optimization approach can be effectively implemented by CFD simulation. Liu et al. [48] studied numerical and experimental studies on a new type of variable rate of the fluidic sprinkler. Zhu et al. [51] carried out the study on orthogonal tests and precipitation estimates for the outside signal fluidic sprinkler. They concluded that the results obtained from the numerical simulation could reflect the inner flow of a complete fluidic sprinkler. Dwomoh et al. [52] compared fluidic and impact sprinklers and concluded that variations in quadrant completion times were small for both fluidic and impact sprinklers. Liu et al. [53] analyzed the droplet size distribution of gas–liquid two phases fluidic sprinkler. Similarly, Yuan et al. [54] studied the effects of a complete fluidic sprinkler on hydraulic characteristics based on some important geometrical parameters. Yuan et al. [55] carried out a study on simulation of combined irrigation for complete fluidic sprinkler based on MATLAB. Li et al. [56] studied the droplet characterization of a complete fluidic sprinkler with different nozzle dimensions. Gan et al. [57] analyzed irrigation uniformity on the complete fluidic sprinkler in a no-wind condition. Hua et al. [58] carried out a study on hydraulic performance of low-pressure sprinkler with special shaped nozzles. Xu et al. [59] researched on double nozzles. Li et al. [60] studied fluidic component of complete fluidic sprinkler.

1.2 Types of Sprinklers Irrigation

Sprinkler irrigation can be divided into two categories. It includes a rotating sprinkler and fixed sprinkler. Rotary sprinklers include impact sprinklers, gear-driven sprinklers, reactive sprinklers and jet sprinklers. Fixed-head sprinklers include most of the spray type currently available. In other words, sprinkler hydraulic performance is a function of the sprinkler's physical features, the geometrical parameters, and environmental conditions. Zhu et al. [61] Compared the outside signal fluidic sprinkler and complete fluidic sprinkler. Sourell et al. [62] studied the performance of rotating spray plate sprinklers indoor experiments. Abo-Ghobar and Al-Amoud [63] analyzed center pivot water application uniformity relative to travel speed and direction. Bishawand, Olumana1958 evaluated the effect of operating pressure and riser height on irrigation water application. Zhu et al. [61] carried a study on the comparison of fluidic and impact sprinklers based on hydraulic performance. Burillo et al. [64] conducted a study on initial drop velocity in a fixed spray plate sprinkler. Ascough and Kiker [65] conducted a study on the effect of irrigation uniformity on irrigation water requirements. Dukes [66] studied the effect of wind speed and pressure on a linear move irrigation system uniformity. Liu et al. [67] modeled the application depth and water distribution of a linearly moving irrigation system. Xiang et al. [68] carried out experiments on air and water suction capability of 30PY impact sprinkler. Dwomoh et al. [52] analyzed sprinkler rotation and water application rate for the newly-designed complete fluidic sprinkler and impact sprinkler. The authors findings revealed that different sprinkler types and sizes exhibit different hydraulic performance characteristics. The operating pressure and nozzle characteristics (nozzle opening size, height, shape, and angle) are the primary factors that control the performance of the sprinkler. They are fundamentally important for developing new sprinkler prototypes because the combination of nozzle size and pressure depends on their values.

1.2.1 Impact Sprinkler

An impact sprinkler was developed by a Glendaola, California, citrus grower in 1933. It is driven in a circular motion by a spring- load arm, pushed back each time it comes into contact with the water stream. One end of the rocker spring is inserted into the rocker frame, the other end is inserted into the spring seat. There are many slots at the upper end of the spring seat. The rotating spring seat can adjust the spring elasticity to change the opening angle of the rocker. In order to adjust the water depth of the diverter conveniently and reduce the friction resistance, the rocker arm usually adopts the suspension structure [69, 70]. The rocker arm consists of a rocker arm body, a rocker arm bushing, and a liquid guide. The diversion plate is composed of a diversion plate and a diversion plate. Under the alternating current of the jet and spring, the rocker arm swings back and forth on the rocker arm axis. Its main function is to drive the nozzle to rotate while adjusting the distribution of water and part of the function of air grinding. The impact sprinkler is driven by a spring arm and pushes backward every time it contacts the water flow. This breaks the flow of water and forms a uniform sprinkler area around the sprinkler (Fig. 1.1).

Fig. 1.1 Impact sprinkler

1.2.2 Complete Fluidic Sprinkler

In 2005, researchers at Jiangsu University in China developed the complete fluidic sprinkler. Its principle of operation is based on a "Coanda effect" to perform the function of rotation. It has the flowing advantages, easy to construct, low loss of energy, low price and reasonable uniformity of water application [63, 64, 71]. Fixed head sprinklers are typically used for watering tall trees and plants. They are designed to function by depending on smooth and grooved cones, deflector plates and slots which produce full or nearly full-circle sprays (Fig. 1.2).

The working theory is based on the theory of the Coanda effect. The working principle of the fluidic sprinkler is as follows: as water is ejected from the nozzle of the main tube into the signal thank, a region of low-pressure forms on both sides at the entry into the main jet flow. Fluid flow from the reversing plastic tubing (left) into the right side, forces the jet to deflect towards the right boundary of the signal thank where it eventually attaches. The air gap between the exit at the right side of the element and the water jet is filled by air, such that the pressures on both sides of the main jet are equal. At the same time, the nozzle receives the signal water on the left edge of the water jet, then the signal water flows in the tube to the inlet signal. Taking out the water from the contact signal into the inlet, the small gap is eventually blocked, forming a low-pressure region. When the pressure difference reaches a certain value, the main jet flows to the right-side attachment wall, Water flows from the diameter into the action zone, and the main jet is ejected from the central circular hole. Under the wall attached condition, due to the bending of the main jet, the signal water nozzle is void and no signal water is received, only air is received. After the signal water in the tube is pumped out, the air enters the water inlet through the tube. The pressures on both sides become equal. Alternate air movement from the signal nozzles and the plate cover account for the stepwise rotation [38, 42, 45].

Fig. 1.2 Fluidic sprinkler

1.2.3 Hand-Move Sprinkler System

The manual transverse device consists of a single length of aluminum tube with a fast connector, which has a riser with a sprinkler mounted at the center or the end. The lateral movement of the hand is often referred to as the hand line. Water is transported to the pipeline at a constant distance through a movable or embedded main pipe with a valve outlet. The system consists of one or more branches. Over the past few years, this irrigation system has been used in areas with larger irrigation areas than any other system and is still used for almost all crop and topographic types. The main disadvantage of the system is high labour intensity and high-pressure requirement (Fig. 1.3).

This system is the basis for the development of all mechanical systems. The typical length of manual lateral movement is 1280 feet (390 m), and the typical spacing of lateral movement is 50 and 60 feet (15 and 18 m). The sprinkler spacing along the side is generally 30 to 40 feet (9 to 12 m), which usually corresponds to the individual length of the pipe. The side is usually made of aluminum, which is easy to move. Some transverse systems use polyethylene (PE) tubes. The mainline system is usually buried underground, with risers extending to the ground every 50 to 200 feet (15 to 60 m) and connected horizontally to the ground. Typical sprinkler operating pressure is 40 to 60 pounds per square inch (270 to 410 kPa). Crops irrigated with manual systems may require deep roots or high soil water holding capacity so that the side does not need to return to a specific location in less than five days, but this depends on the availability of sprinkler pipes, labor, and water supply. The distribution uniformity is improved by using alternating devices. In each other irrigation cycle, pipes are placed in the middle. Hand-operated portable systems change from wheel-line and lateral systems. This type of sprinkler system has a lower initial cost but higher labour demand [72].

Fig. 1.3 Hand-move lateral

Fig. 1.4 Permanent system sprinkler equipment and design

1.2.4 Solid Set and Permanent Systems

Water sprinklers are irrigated at fixed locations. The concept of a solid device system is similar to that of a manual horizontal sprinkler system, except that enough horizontal devices are placed in the field so that no moving pipes are needed during the season. The side door is controlled by a valve, which can guide water into the side door for irrigation at any time. Before the end of the irrigation season, the flanks of the solid irrigation system move into the farmland at the beginning of the season. Solid plant systems use labor at the beginning and end of the irrigation season, but minimize labor demand during the irrigation season [73]. Figure 1.4 shows that the permanent system is a holistic system, usually made of PVC plastic pipes, with the main water supply line and sprinkler side permanently buried and left in place. Stationary or solid systems are similar to portable systems. In addition to mainline and lateral holding in a permanent position or growing season, swivel sprayer is becoming the most commonly used sprayer type in fixed systems. Fixed systems are well adapted to crops requiring a frequent application of water to promote germination and frost resistance [74]. Compared with portable systems, solid-state systems have a higher initial investment but require only a small amount of operating labor.

1.3 Sprinkler Hydraulic Performance Parameters

The hydraulic performance of sprinklers is usually classified as sprinkler, overlapping uniformity and droplet size distribution. It is the physical characteristics of sprinkler, nozzle configuration, working pressure, sprinkler spacing and environmental conditions (wind speed and direction). In other words, the hydraulic performance of the sprinkler is a function of its physical characteristics, geometric parameters, and environmental conditions [75–81]. Therefore, the different types and sizes of sprinklers have different hydraulic performance characteristics. Working pressure and nozzle characteristics (nozzle opening size, height, shape, and angle) are the main factors controlling sprinkler performance. The hydraulic performance of the sprinkler is described by its flow rate, wetting radius, distribution pattern, spraying rate and drip diameter [82]. Several experimental studies have been conducted to analyze the drop size distribution using different methods and techniques with different types of sprinklers over the years. Kohl and DeBoer [83] studied droplet size characterization using photographic methods. Kincaid et al. [84], King et al. [85], Kohl et al. [86] analyzed droplet size distribution using optical laser equipment. Their works revealed that the main problems with the laser method are due to coincidence and edge effect errors. Coincidence error occurs when multiple drops pass through the laser beam simultaneously and thus project overlapping shadows on the photodetector. The edge effect occurs when only a fraction of a drop passes through the laser beam along with one of the edges. Even though these problems cannot be eliminated, they can be minimized by validating each drop measurement based on measured drop velocity. Drops with a measured velocity that significantly differs from the terminal velocity of the measured drop size could be attributed to coincidence or edge effect and thus could be discarded from the data set. Chan and Wallander [87] studied the droplet size distribution. The author confirmed that droplet size distribution varies with distance from the sprinkler. However, Knowledge of the drop size distribution is important because they determine the response of sprinkler droplets to wind influence, evaporation and impact on the soil surface. Li [88] analyzed the water application rate. DeBor et al. [89] studied the application of uniformity of a sprinkler as an important performance criterion for designed and evaluating of sprinklers. The researchers work demonstrated that application rate depends on the operating pressures, the nozzle size and distance between sprinklers. However, the effect of operating pressure on the application rate is minimal compared to the effect of sprinkler nozzle on application rate. For most sprinklers, when operating pressure increases, the discharge tends to be offset by the increase in wetted area. It was found that sprinkler nozzle that produces little droplets covers a smaller wetted area and also have the highest average application rate. Increasing nozzle diameter usually increases the average application rate, since the sprinkler discharge tends to increase more rapidly than the wetted area. Numerous other researchers have proposed an equation to express the coefficient of uniformity. All the above coefficients of uniformity are based on some measures of variation in water distribution. Kay [90], Keller and Bliesner [91] considered a coefficient of uniformity value of less than 85% as low and CU of 85% or above is desirable. According to Zhu et al. [42] Christiansen's coefficient of uniformity is the most widely used and accepted criterion.

1.3.1 Sprinkler Discharge

Sprinkler irrigation discharge is one of the indexes of sprinkler irrigation hydraulic performance, which refers to the amount of water flowing out from the sprinkler irrigation machine per unit time. The main factors affecting the nozzle flow rate are the effective pressure of the nozzle and the effective diameter of the nozzle diameter. The relationship between nozzle discharge and hydraulic parameters is generally expressed as follows:

$$C = \frac{Q}{A\sqrt{2gH}} \tag{1.1}$$

where q is the volumetric discharge of the sprinkler ($m^3\ s^{-1}$), A is the nominal cross-sectional area of the nozzle (m^2), g is the gravitational acceleration ($m\ s^{-2}$), H is the pressure head (m), c is the discharge coefficient, and the discharge exponent for the sprinklers was 0.5.

Since the combination of nozzle size and pressure depends on their values [61], they are essential for the development of new sprinkler prototypes. However, in the field of irrigation, the discharge of a single sprinkler in a given system depends on the depth of water that must be applied for each irrigation, the representative area of the sprinkler, and the water consumption time of a single device, as follows:

$$qs = \frac{d_g S_{L S_M}}{3600 T_S} \tag{1.2}$$

where q_S =, sprinkler discharge, Ls^{-1}; S_L = sprinkler spacing, m; S_M = transverse spacing, m; T_S = operational time per set for signal lateral, hours.

1.3.2 Patterns Radius (Throwing Distance)

The radius of flow pattern is another important hydraulic performance parameter that determines the wet area, average application rate, and runoff potential of the basin. The model radius is defined as the distance from the sprinkler centerline to the farthest point where the sprinkler deposits water at a minimum rate of 0.26 mm/h. The injection distance is approximately proportional to the wetting area [92]. Throwing distance, also known as wetting radius, affects sprinkler and transverse spacing, thereby affecting the cost of sprinkler systems [93–95]. The longer the throwing distance is, the larger the sprinkler spacing is, and vice versa. The infiltration radius is determined by the working pressure, the size and shape of the nozzle and the nozzle axis above the level. Initially, as the working pressure increases, the infiltration radius increases, and then becomes more uniform. To fully break up the water jet, the pressure should increase with the increase of nozzle size. Compared with the nozzles with larger droplets, the nozzles with smaller droplet sizes tend to have shorter injection distances. To study the effects of various operating parameters and sprinkler components on the shape and radius of the model, extensive studies have been carried out [96] used special nozzles with internal contraction angles ranging from 200 to 900 to study the effect of nozzle internal contraction angle on nozzle performance, emphasizing that the flow coefficient and mode radius of the nozzle decrease with the increase of the reverse angle. The action-angle, especially when the contraction exceeds 600. The research also shows that the nozzle contraction angle is suitable for forming uniform water distribution within a larger radius of the pattern, which is about 300. The pattern radius of the nozzle is closely related to the flow coefficient.

A smaller flow coefficient usually results in a shorter graph radius [97] studied the sprinkler spacing of rotating plates on the flanks of continuous mobile irrigation. Their research was carried out in a laboratory at the University of South Dakota, which has two rotating plate spray rotors and rotators manufactured by Nelson Manufacturing Company. In their investigation, nozzle pressures of 50 to 200 kPa and diameters of 4.8, 6.4, 7.9 and 9.9 mm were evaluated. The water application mode of sprinkler under windless conditions was measured in the laboratory. The sprinkler was located 2.5 m above the collection container. The results showed that the wet radius of the rotating nozzle was larger than that of the rotating nozzle under the same nozzle pressure and diameter. The wetting radius of the rotary sprinkler is 1 m (12%) and larger than 4 and 6-groove sprinkler plates. The sprinkler has 100 kPa nozzle pressure and a 6.4 mm in diameter. They also concluded that changes in nozzle diameter would not result in significant differences in the wetting radius and maximum application rate at the far end of the wetting mode, but changes in nozzle pressure would result. They believe that the spraying rate varies greatly with distance, and the relative shape of wetting mode will affect the uniformity of water spraying.

1.3.3 Water Application Rate or Intensity

The design application efficiency is based on the expected potential performance of the sprinkler system, before installation, based on an analysis of a proposed system layout and configuration. But the calculated design application efficiency also presupposes correct operation (pressures, set durations, and other factors) and maintenance of the system. In contrast, the actual application efficiency is measured in the field on an existing sprinkler system for comparisons and the identification of changes that could be made to improve the system performance. Application rate depends on the operating pressures, the nozzle size and distance between sprinklers [98]. However, the effect of operating pressure on the application rate is minimal compared to the effect of a sprinkler nozzle on the application rate. For most sprinklers, when operating pressure increases, the discharge tends to be offset by the increase in wetted area. It has been found that sprinkler nozzle that produces little droplets covers a smaller wetted area and also has the highest average application rate. Increasing nozzle diameter usually increases the average application rate, since the sprinkler discharge tends to increase more rapidly than the wetted area [99]. The water application rate is the primary performance variable on which most sprinkler performance indicators are derived. It is usually measured experimentally by indoor or outdoor catch-can (rain gauge) experiments [100–102].

The rate at which water should be applied depends on: The time required for the soil to absorb the calculated depth of application without runoff for the given conditions of soil, slope, and cover. The depth of application divided by this required time is the maximum application rate.

The minimum application rate that will result in uniform distribution and satisfactory efficiency under prevalent climatic conditions or that is practical with the system selected. For most irrigated crops, the minimum rate of application to obtain reasonably good distribution and high efficiency under favorable climatic conditions is about 0.15 inch per hour (4 mm/h). If high temperatures and high wind velocities are common, the minimum application rate must be higher to reduce problems associated with wind drift. The establishment of minimum application rates for local conditions requires both experience and judgment. The amount of time it takes for irrigation to achieve efficient use of available labor in coordination with other operations on the farm. The application rate adjusted to the number of sprinklers operating in the best practice system layout. The application rate is calculated based on the weighted average of the area represented by each measuring point using the equation

$$A = K\frac{Q}{LS} \tag{1.3}$$

where, A—is the application rate in mm/hr; Q—is the sprinkler discharge in l/s; L—is the lateral spacing in m; S—is the sprinkler spacing on a lateral in m; K—is a conversion constant. In the case of a single sprinkler, the application rate is calculated by replacing LS with a—the wetted area of a sprinkler in m^2.

1.3.4 Distribution Pattern

The distribution mode can be defined as the change mode of water quantity and spray quantity of sprinklers. It is also known as water distribution or precipitation profile [103]. The distribution pattern is a function of many factors, i.e. nozzle pressure, nozzle shape and size, nozzle design, presence of straightening blades, nozzle speed, trajectory angle, riser height, and wind. The radial water distribution of a sprinkler with a given pressure combination can be easily and effectively measured and expressed as a function without wind control [104, 105]

$$P_i = f_i(r) \tag{1.4}$$

where P = the rate of water application at distance r from the sprinkler; and i = indoor conditions. Water distribution patterns are characteristic of the sprinkler at a given pressure.

Nozzle characteristics do not affect the distribution pattern as much as the operating pressure. These patterns have different shapes that are distinct from sprinkler to sprinkler.

1.3.5 Sprinkler Droplet Size

Droplet size characteristics have been used for different purposes related to irrigation management, such as evaporation loss, soil conservation, and irrigation simulation. The drop diameter, velocity, and trajectory of soil surface depend on many factors. The most relevant are the types of sprinklers and nozzles, the operating hydraulic parameters and the environmental conditions of the specific location of the sprinkler system. Ballistic theory constitutes the most common modeling method for sprinkler irrigation, especially for solid solidification systems [106]. Droplet size characteristics are very important in sprinkler irrigation for number reasons. Firstly, wind upsets the uniformity and efficiency of sprinkler systems by distorting the water distribution pattern of sprinkler [107]. The degree of distortion depends on the wind speed and direction as well as the size of droplets in the distribution. The force exerted by the wind is proportional to the square of the droplet diameter, while the inertia of the droplet resisting the wind is proportional to the mass of the droplet (the function of the cubic diameter of the droplet). The wind tends to distort the distribution of the droplet in sprinkler irrigation, while the square of the diameter of the droplet in sprinkler irrigation is proportional to the inertia of the droplet. Secondly, during sprinkler irrigation, besides climate demand, the amount of water evaporated by water droplets depends on the time and surface area of water droplets in the air [96]. The time that water droplets stay in the air and the surface area of water droplets are both functions of the size of water droplets. For water per unit volume, the surface area is doubled because the diameter of water droplets is reduced by half. As the evaporation rate increases with the increase of the exposed area, when other factors remain unchanged, the evaporation rate increases with the decrease of droplet size. The available time for droplet evaporation is from the time the droplet leaves the nozzle to the time it falls on the ground or the surface of the crop. When this time is prolonged sufficiently by wind-suspended droplets, small droplets will evaporate before landing [68]. The extent of wind drift is also affected by the size of water droplets [108–110]. The third-largest water drop may cause soil erosion, which is due to the decrease in water permeation rate. This is due to the high kinetic energy of large droplets, which can destroy the soil surface, especially in crust-affected soils, resulting in the sealing of the soil surface [111, 112]. As a result, it is becoming more and more convenient for irrigators to reduce runoff and erosion by converting sprinklers that emit large amounts of water droplets into sprinklers that emit smaller ones [113, 114]. Studies show that droplet formation is mainly influenced by pressure, nozzle size, and structure [115, 116]. Operating the sprinkler under higher pressure can increase the percentage of small water droplets and reduce the percentage of large water droplets. In recent years, sprinkler irrigation systems operating at lower pressure have attracted much attention due to the rising energy costs. However, [117], operating a sprinkler at lower pressure usually means larger droplets and uneven distribution patterns, especially for established systems (with circular nozzles). With such a system, the inefficiency of applications is inevitable.

However, some researchers have shown that by oscillating pressure centered on a relatively low average value.

1.3.6 Sprinkler Irrigation Uniformity

The purpose of sprinkler irrigation is to distribute water to a predetermined area in the form of absorption and uniform depth so that each part of the irrigation area can obtain the same amount of water [118–120]. This is an ideal situation, it is impossible, even if there is natural rainfall [52]. The term irrigation uniformity refers to the change or non-uniformity of water consumption of irrigation area. Irrigation uniformity is an important aspect of system performance. This is a common indicator to measure the efficiency of agricultural technology, which measures the spatial change of water use [53]. The uniformity of irrigation is the main factor that affects the normal growth of crops. So far, the uniformity of sprinkler irrigation has been the focus of sprinkler system design and management [61, 121]. A specific quantitative study of sprinkler irrigation study started with the pioneering work of Christiansen [111]. Parameters that are used to evaluate sprinkler irrigation uniformity are the coefficient of uniformity and distribution uniformity. The performance is often evaluated based on water uniformity coefficients collected in an array of rain gauges. A design uniformity of CU less than 95 percent is not normally warranted for any type of sprinkler system because the increased system cost may be more than the benefits of such a high uniformity. Thus, the target uniformity for system design should not be unreasonably high. Also, it should be recognized that the effective uniformity of sprinkler irrigation is usually higher than the application uniformity, at least in the absence of ponding and runoff, because of spatial water redistribution within the crop root zone, both during and immediately following irrigation. Some of the factors that affect uniformity tend to average out during multiple irrigation applications.

Table 1.1 presents the recommended minimum CU values of the different types of sprinkler irrigation systems. In general, economic and production criteria suggest a target CU of at least 85 percent for delicate and shallow-rooted crops such as potatoes and most other vegetables. Deep-rooted field crops such as alfalfa, corn, cotton, and sugar beets and tree and vine crops that have deep spreading root systems can generally be adequately irrigated using the values also listed in Table 1.1 [105].

Other researchers have proposed an equation to express the coefficient of uniformity. All the above coefficients of uniformity are based on some measures of variation in water distribution [107, 96]. A coefficient of uniformity value of less than 85% as "low" and CU of 85% or above as "desirable". Christiansen's coefficient of uniformity is the most widely used and accepted criterion [112–114].

$$CU = 100\left(1 - \frac{\sum_{i=1}^{n} |X_i - \mu|}{\sum_{i=1}^{n} X_i}\right) \tag{1.5}$$

Table 1.1 Recommended design CU values for different sprinkler systems

Sprinkler type	Recommended minimum design CU (%)
Hand move	70
Side roll	75
Fixed (solid set)	85
Traveler	85
Center pivot	90
Linear move	95
Disposal of effluent	70

where n = number of catch cans; x_i = measured application depth, mm; μ = mean average depth, mm; and CU = coefficient off uniformity, %

Combined CU developed by Christiansen [111]

$$CU = 100\left(1 - \sum_{i=1}^{n}\left|h_i - \bar{h}\right| \Big/ n\,\bar{h}\right) \quad (1.6)$$

where h_i = water depth of calculated point, mm h", h = mean water depth of al calculated point within the area, mm h"; and the n = total number of calculated points used in the equation.

Christiansen uniformity coefficient

$$CU = 100 \times \left(1 - \frac{\sum X}{n \cdot m}\right) \quad (1.7)$$

where CU is the coefficient of uniformity Christiansen (%), $\sum$x is the summation of deviation from the mean depth collected; m is the mean depth collected and n is the number of observations.

Christiansen

$$CU_c = \left(1 - \frac{\sum_{i=1}^{n} abs(x_i - x)}{Nx}\right) \times 100 \quad (1.8)$$

where X = mean water collected depth, mm; Xi is the water collected depth, mm: is the sum of the absolute deviation from the mean, X of all N observations mm;

Christiansen CU

$$\mathrm{CU} = 100\left(1.0 - \frac{x}{mn}\right) \quad (1.9)$$

where, x is the sum of the deviation of each observation from m, the mean value of such observations and N is the number of observations.

References [113–116] also formulated their coefficient similar to that of Christiansen [111]. However, [114] limited their equation to the lowest quarter depths of water application, whereas [210] limited the equation to the highest quarter depths off water application.

$$CU = 100\left(1 - \frac{\sum_{i=1}^{\frac{n}{4}} |X_i - \mu|}{\mu \times \frac{n}{4}}\right) \quad (1.10)$$

$$CU = 100\left(1 - \frac{\sum_{i=\frac{3}{4}n+1}^{n} |X_i - \mu|}{\mu \times \frac{n}{4}}\right) \quad (1.11)$$

Their uniformity was based on the squares of the deviations from the mean instead of the deviation themselves, contrarily to that of Christiansen.

$$\mathrm{CU} = 100\left(1 - \frac{\sigma}{\mu}\right) \quad (1.12)$$

References [111–114] also proposed distribution efficiency, which is based on the numerical integration of the normal distribution function. The approach requires first selecting a target CU and a target percentage area, which will be adequately irrigated.

$$\sum_{i=}^{n} |X_i - \mu| \cong n\sigma\sqrt{2/\pi} \quad (1.13)$$

$$\mathrm{CU} = 100\left(1 - \frac{0.798\sigma}{\mu}\right) \quad (1.14)$$

where σ = standard deviation of all depth measurements, mm. Where n = number of catch cans; x_i = measured application depth, mm; μ = mean average depth, mm; and CU = coefficient off uniformity.

1.3.7 Methods of Measuring Droplet Size Distributions

Many methods have been applied to the characterization of spray droplets, such as dyeing, oil immersion, momentum, photography or optical methods. These techniques were used in precipitation studies [67, 97, 106, 122], some of which were later used to assess sprinkler irrigation. This is the case of [123], who implemented the flour method. In this method, droplets impinge on the thin flour layer to form particles whose mass is related to the diameter of droplets [124] proposed a technique based on image processing to measure droplet size by optical spectrophotometer. Recently, [125] used low-speed photography to characterize the characteristics of water droplets at different horizontal distances and impact sprinklers. At the end of the twentieth century, several techniques were developed to measure the flow of water under different hydraulic conditions. These technologies include laser dropper velocimetry (LDA), hotline velocimetry (HWA), laser-induced fluorescence (LIF), dropper global velocimetry (DGV) and Doppler phase velocimetry (PDA). Due to the emergence of high-speed cameras and advanced laser equipment, a non-invasive technique for evaluating flow characteristics is proposed by using computational visualization methods such as two-dimensional particle image velocimetry (PIV) and particle tracking velocimetry (PTV) for two-dimension analysis.

References

1. Hussein M, Abo-Ghobar K (1994) The effect of riser height and nozzle size on evaporation and drift losses t inner arid conditions [J]. J King Saud Univ Agric Sci 2:191–202
2. Oweis T, Hachum A (2006) Water harvesting and supplemental irrigation for improved water productivity of dry farming systems in West Asia and North Africa [J]. Agric Water Manag 80:57–73
3. Rosenzweig C, Strzepek K, Major D, Iglesias A et al (2004) Water availability for agriculture under climate change: five international studies [J]. Glob Environ Change 14:345–360
4. Sadiddin A (2013) An assessment of policy impact on agricultural water use in the northeast of Syria [J]. Environ Manag Sustain Dev 2:74–105
5. Iglesias A, Quiroga S, Diz A (2011) Looking into the future of agriculture in a changing climate.[J] Eur Rev Agric Econ 38(3):427–447
6. Yigezu YA, Aw-Hassan A, Shideed K et al (2014) A policy option for valuing irrigation water in the dry areas[J]. Water Policy 16:520–535
7. Beaumont P (1996) Agricultural and environmental changes in the upper Euphrates catchment of Turkey and Syria and their political and economic implications[J]. Appl Geogr 16:137–157
8. Shah T, Roy AD, Qureshi AS et al (2003) Sustaining Asia's groundwater boom an overview of issues and evidence [J]. Sustain Dev J 27(2):130–141
9. Soussan J, Arriens WL (2004) Poverty, and water security understanding how water affects the poor water for all series No. 2 [M]. Asian Development Bank (ADB)
10. Schlenker W, Lobell D (2010) Robust negative impacts of climate change on African agriculture [J]. Environ Res Lett 5(1):1–10
11. UNECE (United Nations Economic Commission for Europe) (2011) Second Assessment of Transboundary Rivers, Lakes, and Groundwater [M]. New York/Geneva, UNECE
12. Kennedy AJ (2016) Determining climate effects on total agricultural productivity [J]. Natl Acad Sci 114(12):2285–2292
13. Kahlown MA, Raoof A, Zubair M et al (2007) Water use efficiency and economic feasibility of growing rice and wheat with sprinkle irrigation in the basin of Pakistan [J]. Agric Water Manag 87:292–298
14. Yigezu YA, Ahmed MA, Shideed K et al (2013) Implications of a shift in irrigation technology on resource use efficiency: a syrian case. Agric Syst 118:14–22
15. Haman Dorota Z, Thomas H (2005) Field evaluation of container nursery irrigation systems [M] uniformity of water application. IFAS FS98-2
16. Dechmi F, Playan E, Faci JM (2003) Analysis of an irrigation district in northeastern Spain characterization and water assessment [J]. Agric Water Manag 61:75–92
17. Dukes MD, Zotarelli L, Morgan KT (2010) Use of irrigation technologies for vegetable crops in Florida [J]. Hort Technol 20(1):133–142
18. Dechmi F, Playa-n E, Cavero J (2004) A coupled crop and solid set sprinkler simulation model Il model application [J]J. Irrig Drain Eng ASCE 130(6):511–519
19. Berkes F (2009) Evolution of co-management: Role of knowledge generation, bridging organizations and social learning [J]. J Environ Manag 90:1692–1702

20. Gielen D, Steduto P, Mueller A et al (2011) Considering the energy, water and food nexus towards an integrated modelling approach [J]. Energy Policy 39:7896–7906
21. Endo A, Tsurita I, Burnett K et al (2017) A review of the current state of research on the water, energy, and food nexus[J]. J Hydrol Reg Stud 11:20–30
22. Pittock J, Hussey K, McGlennon S (2013) Australian climate energy and water policies conflicts and synergies[J]. Aust Geogr 44:3–22
23. Biggs EM, Bruce E, Boruff B et al (2015) Sustainable development and the water energy food nexus perspective on livelihoods [J]. Environ Sci Policy 54:389–397
24. Rodell M, Famiglietti JS, Wiese DN et al (2018) Emerging trends in global freshwater availability [J]. Nature 557:651–659
25. Liu J, Hull V, Godfray HCJ et al (2018) Nexus approaches to global sustainable development [J]. Nat Sustain 1:466–476
26. Cairns R, Krzywoszynska A (2016) Anatomy of a buzzword the emergence of the water-energy food nexus' in UK natural resource debates [J]. Environ Sci Policy 64:164–170
27. Leck H, Conway D, Bradshaw M et al (2015) Tracing the water–energy–food nexus description, theory and practice [J]. Geogr Compass 9:445–460
28. Momblanch A, Papadimitriou L, Jain SK et al (2019) Untangling the water-food-energy environment nexus for global change adaptation in a complex himalayan water resource system [J]. Sci Total Environ 655:35–47
29. Allen C, Metternicht G, Wiedmann T (2018) Initial progress in implementing the Sustainable Development Goals (SDGs) a review of evidence from countries [J]. Sustain Sci 13:1453–1467
30. Chen C, Park T, Wang X et al (2019) China and India lead in greening of the world through land-use management [J]. Nat Sustain 2:122–129
31. Bryan BA, Gao L, Ye Y et al (2018) China's response to a national land-system sustainability emergency[J]. Nature 559:193–204
32. Chen X (2009) Review of China's agricultural and rural development: Policy changes and current issues[J]. China Agric Econ Rev 1:121–135
33. Snilstveit B, Oliver S, Vojtkova M et al (2012) Narrative approaches to systematic review and synthesis of evidence for international development policy and practice [J]. J Dev Eff 4:409–429
34. Mwangi M, Kariuki S (2015) Factors determining adoption of new agricultural technology by small holder farmers in developing countries [J]. J Econ Sustain Dev 6:208–216
35. Wang J, Rothausen SG, Conway D et al (2012) China's water energy nexus: greenhouse-gas emissions from groundwater use for agriculture [J]. Environ Res Lett 7:14035
36. Chen D, Yu Q, Hu Q et al (2018) Cultivated land change in the belt and road initiative region [J]. Geogr Sci 28:1580–1594
37. Seginer I, Kantz D, Nir D et al (1992) Indoor measurement of single-radius sprinkler patterns [J]. Trans ASAE 35
38. Gan ZM, Chen S (1980) Self-controled complete fluidic sprinkler going forward step-by-step. Chinese patent Office [P]. Patent No. 87203621.2404
39. Zhu XY, Jiang JY, Liu JP et al (2015) Compared between outside signal fluidic sprinkler and complete fluidic sprinkler. J Drain Irrig Mach Eng 33(2):172–178
40. Dwomoh FA, Yuan SQ, Li H (2013) Field performance characteristics of fluidic sprinkler[J]. Appl Eng Agric 29(4):529–536
41. Zhang L, Merkley GP, Pinthong K (2013) Assessing whole-field sprinkler irrigation application uniformity[J]. Irrig Sci 31(2):87–105
42. Zhu X, Yuan S, Liu J (2012) Effect of sprinkler head geometrical parameters on hydraulic Performance of fluidic sprinkler [J]. J Irrig Drain Eng 138:1019–1026
43. Li H, Yuan SQ, Xie FQ et al (2006) Performance characteristics of fluidic sprinkler controlled by clearance and comparison with impact sprinkler[J]. Trans CSAE 22(5):82–85
44. Li H, Yuan SQ, Liu JP et al (2007) Wall-attachment fluidic sprinkler. Ch. Patent No. 101224444 B
45. Yuan SQ, Zhu XY, Li H (2005) Numerical simulation of inner flow for complete fluidic sprinkler using computational fluid dynamics[J]. Trans CSAM 36(10):46–49

46. Li H, Yuan SQ, Xiang QJ et al (2011) Theoretical and experimental study on water offset flow in fluidic component of fluidic sprinklers[J]. J Irrig Drain Eng 137(4):234–243
47. Hu G, Zhu XY, Yuan SQ et al (2019) Comparison of ranges of fluidic sprinkler predicted with BP and RBF neural network models[J]. J Drain Irrig Mach Eng 37(3):263–269
48. Liu J, Zang C, Tian S et al (2013) Water conservancy projects in China: achievements, challenges and way forward [J]. Glob Environ Chang 23:633–643
49. Xu SR, Wang XK, Xiao SQ, Fan ED (2019) Numerical simulation and optimization of structural parameters of the jet-pulse tee[J]. J Drain Irrig Mach Eng 37(3):270–276
50. Song HB, Yoon SH, Lee DH (2000) Flow and heat transfer characteristic of a two dimensional oblique wall attaching offset jet[J]. Int J Heat Mass Transf 43:2395–2398
51. Zhu X, Yuan Shouqi, Li Hong (2009) Orthogonal test for the structure parameters of complete fluidic sprinkler [J]. Trans CSAE 25:103–107
52. Dwomoh FA, Yuan S, Li H (2014) Droplet size characterization of the new type complete Fluidic sprinkler. J Mech Civil Eng 11:70–73
53. Liu JP, Liu WZ, Bao Y et al (2017) Drop size distribution experiments of gas-liquid two phase's fluidic sprinkler [J]. J Drain Irrig Mach Eng 35:731–736
54. Yuan S, Zhu X, Li H et al (2006) Effects of complete fluidic sprinkler on hydraulic characteristics based on some important geometrical parameters[J]. Trans CSAE 22(10):113–116
55. Yuan S, Zhu XY, Hong L (2008) Simulation of combined irrigation for complete fluidic sprinkler based on MATLAB [J]. J Drain Irrig Mach 26(01):47–52
56. Li J, Kawano H, Yu K (1994) Droplet size distributions from different shaped sprinkler nozzles[J]. Trans ASAE 37(6):1871–1878
57. Gan ZM (1985) Analysis on complete fluidic sprinkler typed PSZ[J]. Trans CSAM 16(6):87–91
58. Hua L, Jiang Y, Li H, Zhou XY (2018) Hydraulic performance of low-pressure sprinkler with special-shaped nozzles [J]. J Drain Irrig Mach Eng 36(11):1109–1114
59. Xu SR, Wang XK, Xiao SQ, Fan ED, Zhang CX, Xue ZL, Wang X (2018) Experimental study on the double-nozzle jet sprinkler [J]. J Drain Irrig Mach Eng 36(10):981–984
60. Li H, Wang C, Chen C, Shen ZH (2013) Study on fluidic component of complete fluidic sprinkler. Adv Mech Eng 10:65–85
61. Zhu XY, Yuan SQ, Jiang JY et al (2015) Comparison of fluidic and impact sprinklers based on hydraulic performance [J]. Irrig Sci 33(5):367–374
62. Sourell H, Faci JM, Playan E (2003) Performance of rotating spray plate sprinklers indoor experiments [J]. J Irrig Drain Eng 129:376–380
63. Abo-Ghobar HM, Al-Amoud AI (1993) Center pivot water application uniformity relative to travel speed and direction [J]. Alex J Agric 38(I):1–18
64. Burillo GS, Delirhasannia R, Playán E (2013) Initial drop velocity in a fixed spray plate sprinkler[J]. J Irrig Drain Eng ASCE 139(7):521–531
65. Ascough GW, Kiker GA (2002) The effect of irrigation uniformity on irrigation water requirements. Water SA 28(2):235–241
66. Dukes MD (2006) Effect of wind speed and pressure on linear move irrigation system uniformity[J]. Appl Eng Agric 22(4):541–548
67. Liu J, Zhu X, Yuan S et al (2019) Modeling the application depth and water distribution of a linearly moving irrigation system[J]. Water 10:1301
68. Xiang QJ, Xu ZD, Chen C (2018) Experiments on air and water suction capability of 30PY impact sprinkler [J]. J Drain Irrig Mach Eng 36:82–87
69. Zoldoske DF (2007) An overview of smart water application technologies and achieving high water use efficiency [M]. In: Proceedings of California soil, and plant conference: opportunities for California agriculture, American society of agronomy, Sacramento, California
70. Wall RW, King BA (2004) Incorporating plug and play technology into measurement and control systems for irrigation management[M]. ASABE Paper No. 042189. St. Joseph, MI: ASABE

71. Bishaw D, Olumama M (2015) Evaluating the effect of operating pressure and riser height on irrigation water application under different wind conditions in Ethiopia[J]. Asia Pac J Energy Environ 2:41–48
72. De Wrachien D, Lorenzini G, Medici M (2013) Food production and irrigation and drainage systems development perspective and challenges[J]. Irrig Drain Syst Eng 2(3)
73. Hsiao TC, Steduto P, Fereres E (2007) A systematic and quantitative approach to improve water use efficiency in agriculture [J]. Irrig Sci 25:209–231
74. Schneider AD, Howekk TA (1999) LEAP and spray irrigation for grain crops [J]. J Irrig Drain Eng 125(4):167–172
75. ASAE American Society of Agricultural Engineers (2001) The test procedure for determining the uniformity of water distribution of center pivot and lateral move irrigation machines equipped with spray or sprinkler nozzles [S]. ASAE Standards, ANSI/ASAE S436.1 MAROI
76. Louie MJ, Selker J (2000) Sprinkler head maintenance effects on water application uniformity [J]. J Irrig Drain Eng ASCE 126(3):142–148
77. Yan HJ. Jin HZ (2004) Study on the discharge coefficient of non-rottable sprays for center-pivot system [J]. J Irrig Drain 23:55–58
78. Oniward S, Hardlee M, Sherman M (2010) An investigation of the energy saving potential of mini- sprinkler irrigation systems [J]. Int J Sci Technol 2(7):3287–3296
79. Frederick RW (2009) Seventy-fifth anniversary of horizontal action impact drive sprinkler [J]. J Irrig Drain Eng 1352:133
80. Sudheera KP, Panda RK (2000) Digital image processing for determining drop sizes from irrigation spray nozzles [J]. Agric Water Manag 45:159–167
81. Smajstrla AG, Boman BJ, Haman DZ (2006) Basic irrigation scheduling in Florida [M]. Florida Coop Ext Serv Inst Food Agric Sci Univ Florida.
82. Yan HJ, Liu ZQ, Wang FX et al (2007) Research and development of impact sprinkler in China.[J]. J China Agric Univ 12:77–80
83. Kohl RA, DeBoer DW (1983) Drop size distributions for a low-pressure spray type agricultural sprinkler [S]. ASAE Paper 83–2019, St. Joseph, MI
84. Kincaid DC (1996) Spray drops kinetic energy from irrigation sprinklers [J]. Trans ASAE 39(3):847–853
85. King BA, Bjorneberg DL (2010) Charactering droplet kinetic energy applied by moving spray-plate center- pivot irrigation sprinklers [J]. Trans ASABE 53:137–145
86. Kohl RA, Bernuth RD, Heubner G (1985) Drop size distribution measurement problems using a laser unit[J]. Trans ASAE 28(1):190–192
87. Chan D, Wallender WW (1985) Droplet size distribution and water application with low pressure sprinkler[J]. Trans ASAE 11:801–803
88. Li JR (2000) Sprinkler water distributions as affected by winter wheat canopy[J]. Irrig Sci 20(1):29–35
89. DeBor DW, Monnens MJ, Kincaid DC (2001) Measurement of sprinkler droplet size [J]. Appl Eng Agric 17:11–115
90. Kay (1988) Sprinkler irrigation equipment and practice [M]. Batsford Limited, London
91. Keller RD, Bliesner RD (1990) Sprinkler Irrigation [M]. Van Nostrand Reinhold, USA, New York
92. Jin HZ, Yan HJ, Qian YC (2010) Overseas development of water saving irrigation engineering technology[J]. Trans ASAME 41(1):59–63
93. Burt CM, Clemmens AJ, Strelkoff KH (1997) Irrigation performance measures: efficiency and uniformity[J]. J Irrig Drain Eng 123(6):423–442
94. Suharto B, Susanawati LD (2012) Design and construction of sprinkler irrigation for stabilizing apple crop in dry season[J]. J Appl Environ Biol Sci 2:134–139
95. King BA, Wall RW, Kincaid DC (2017) Field testing of a variable rate sprinkler and control system for site-specific water and nutrient application[J]. Trans ASAE 21(5):847–853
96. Liu JP, Liu XF, Zhu XY et al (2016) Droplet characterization of a complete fluidic sprinkler with different nozzle dimensions [J]. Biosyst Eng 65(4):2–529

97. Salles C, Poesen J, Borselli L (2012) Measurement of simulated drop size distribution with an optical spectro pluviometer sample size concentration [J]. Earth Surf Process Landforms 24(6):5–45
98. Duan FY, Liu JR, Fan YS et al (2017) Influential factor analysis of spraying effect of light hosefed traveling sprinkling system [J]. J Drain Irrig Mach Eng 35(6):541–546
99. Pan Y, Suga K (2003) Capturing the pinch-off of the liquid jets by the level set method [J]. J Fluids Eng 125:922–930
100. Negeed SR, Hidaka S, Kohno M (2009) Experimental and analytical investigation of liquid sheet breakup characteristics [J]. Int J Heat Fluid Flow 32(1):95–106
101. Xu D, Li YN, Gong SH et al (2019) Experiment on sweet pepper nitrogen detection based on near-infrared reflectivity spectral ridge regression [J]. J Drain Irrig Mach Eng 37(1):63–72
102. Li YF. Liu JP, Li T et al (2018) Theoretical model and experiment on fluidic sprinkler wet radius under multi-factor [J]. J Drain Irrig Mach Eng 36(8):685–689
103. Sadeghi SH, Peters T, Shafii B et al (2017) Continuous variation of wind drift and evaporation losses under a linear move irrigation system [J]. Agric Water Manag 182(3):39–54
104. Coanda H (1936) Device for deflecting a stream of elastic fluid projected into an elastic fluid [P]. U.S. Patent No. 2,052,869
105. Zhu XY, Peters T, Neibling H (2016) Hydraulic performance assessment of LESA at low pressure [J]. Irrig Drain 65(4):530–536
106. Tang LD, Yuan SQ, Qiu ZP (2018) Development and research status of water turbine for hosereel irrigator[J]. J Drain Irrig Mach Eng 36(10):963–968
107. Li J (2000) Sprinkler irrigation hydraulic performance and crop yield [D]. Thesis Report. ISBN 780119888-3. Chapter 6: pp 75–85. Kagawa University, Japan
108. Erdal G, Esengun K, Erdal H et al (2007) Energy use and economic analysis of sugar beet production in Tokat Province of Turkey[J]. Energy 32:35–41
109. Lopez-Mata E, Tarjuelo JM, Juan Ballesteros JA et al (2010) Effect of irrigation uniformity on the profitability of crops[J]. Agric Water Manag 98:190–198
110. Wang G, Wang W, Xu F (2006) The optimum calculation of the combination of sprinkler and the combination spacing of the sprinkler system [J]. J Heilongjiang 33(2):36–39
111. Christiansen JE (1942) Irrigation by sprinkling. California agricultural experiment station [M]. Bulletin, vol 670. University of California, Berkeley, CA
112. Criddle WD, Davis S, Pair CH, et al (1956) Methods of evaluating irrigation systems. In: Agriculture handbook No.82 soil conservation service [M]. USDA, Washington, DC
113. Maroufpour E, Faryabi A, Ghamarnia H et al (2010) Evaluation of uniformity coefficients for sprinkler irrigation systems under different field conditions in Kurdistan Province northwest of Iran[J]. Soil Water Res 5(4):139–145
114. Wilcox JC, Swailes GE (1947) Uniformity of water distribution by some under tree orchard sprinklers [J]. J Sci Agric 27:565–583
115. Han WT, Fen H, Yang Q et al (2007) Evaluation of sprinkler irrigation uniformity by double interpolation using cubic splines[M]. In: Effective utilization of agricultural soil & water resources and protection of environment, pp 250–255
116. Thooyamani KP, Norum DI, Dubetz S (1984) Spray patterns shape and application rate patterns under low-pressure sprinklers [S]. ASAE paper No. 84-2588. St Joseph, Mich.: ASAE
117. Lan CY, Yi XT, Xue GN (2005) Research state and development of sprinkler irrigation equipment in China[J]. Dain Irrig Mach 23(1):1–6
118. Wei Y (1996) Fiscal systems and uneven regional development in China, 1978–1991 [J]. Geoforum 27:329–344
119. Oksanen T, Ohman M, Miettinen M et al (2004) Open configurable control system for precision farming. ASABE Paper No. 701P1004. St. Joseph, MI: ASABE
120. Ali OOA (2008) Simulation model for centre pivot system design and optimization of operation. PhD. Thesis, University of Khartoum Sudan, pp 1–40
121. Hendawi M, Molle B, Folton C et al (2005) Measurement accuracy analysis of sprinkler irrigation rainfall in relation to collector shape [J]. J Irrig Drain Eng 131:477–483

122. Zhu XY, Chikangaise P, Shi WD et al (2018) Review of intelligent sprinkler irrigation technologies for remote autonomous system. Int J Agric Biol Eng 11(1):23–30
123. Alexander F, Xingye Z, Shouqi Y et al (2020) Numerical simulation and experimental study on internal flow characteristic in the dynamic fluidic sprinkler[J]. Am Soc Agric Biol Eng 36(1):61–70
124. Zhu X, Fordjour A, Yuan S et al (2018) Evaluation of hydraulic performance characteristics of a newly designed dynamic fluidic sprinkler [J]. Water 10:1301
125. Baum MC, Dukes MD, Miller GL (2005) Analysis of residential irrigation distribution uniformity[J]. J Irrig Drain Eng 131(4):336–341
126. Yu J, Wu J (2018) The sustainability of agricultural development in China the agriculture environment nexus [J]. Sustainability 10:1776
127. Liu JP, Yuan SQ, Li H et al (2008) Analysis and experiment for range and spraying uniformity influencing factor of complete fluidic sprinkler. Trans Chin Soc Agric Mach 39(11):51–54
128. Liu JP, Yuan SQ, Li H et al (2013) Numerical simulation and experimental study on a new type variable-rate fluidic sprinkler. J Agric Sci Technol 15(3):569–581
129. Liu J, Yuan S, Li H et al (2013) Theoretical and experimental study of a variable-rate complete fluidic sprinkler [J]. Appl Eng Agric 29(1):17–24
130. Kranz, WL, Suat I, et al (2007) Extension irrigation specialists flow control devices for center pivot irrigation systems [M]. NebGuide Published by University of Nebraska – Lincoln extension, institute of agriculture and natural resources G8888
131. Kincaid DC (1982) Sprinkler pattern radius [J]. Trans ASAE 25:1668–1672
132. Chi. Irrigation and Drainage [M] (2010) Chapter 3 Irrigation Techniques 41–65. ISBN 9787-50847279-9
133. Topak R, Suheri S, Ciftci N et al (2005) Performance evaluation of sprinkler irrigation in a semi arid area [J]. Pak J Bio Sci 8(1):97–103
134. Xiang Q (2008) Study on hydraulic characteristics of the Coanda effect in fluidic sprinkler [D]. Jiangsu University, China
135. He X, Yang PL, Ren SM et al (2016) Quantitative response of oil sunflower yield to evapotranspiration and soil salinity with saline water irrigation [J]. Int J Agric Biol Eng 9(2):63–73
136. Oztan M, Axelrod M (2011) Sustainable transboundary groundwater management under shifting political scenarios: the Ceylanpinar Aquifer and Turkey-Syria relations[J]. Water Int 36:671–685
137. Cao H, Guo FT, Fan YS et al (2016) Running speed and pressure head loss of the light and small sprinkler irrigation system [J]. J Drain Irrig Mach Eng 34(2):179–184
138. Lorenzini G, Wrachien D (2005) Performance assessment of sprinkler irrigation systems a new indicator for spray evaporation losses [J]. Irrig Drain 54(3):295–305
139. Thompson AL, Martin DL, Norman JM (1997) Testing of a water loss distribution model for moving sprinkler systems[J]. Trans ASAE 40(1):81–88
140. Howell TA (2001) Enhancing water use efficiency in irrigated agriculture[J]. Agron J 93(2):281–289
141. Hathoot HM, Abo- HM, Al- AI (1994) Analysis and design of sprinkler irrigation laterals[J]. J Irrig Drain Eng 120(3):534–549
142. Gilley JR, Watts DG (1977) Possible energy saving in irrigation[J]. J Irrig Drain Div ASCE 103(4):445–455
143. DeBoer DW, Monnens MJ, Kincaid DC (2000) Rotating-plate sprinkler spacing on continuous irrigation laterals. National irrigation symposium [R]. In: Proceedings of the 4th decennial symposium, USA, pp 115–122
144. Kara T, Ekmekci E, Apan M (2008) Determining the uniformity coefficient and water distribution characteristics of some sprinklers [J]. Pak J Biol Sci 11:214–219
145. Hsiao TC, Steduto P, Fereres E (2007) A systematic and quantitative approach to improve water use efficiency in agriculture [J]. Irrig Sci 25:209–231
146. Han X, Hao P (2005) Analysis of spraying mechanism of a whirl sprinkler with square spray field[J]. Trans CSAM 3:40–44

147. Zanon ER, Testezlaf R, Matsura EJ (2000) Data acquisition system for sprinkler uniformity testing [J]. Appl Eng Agric 16(2):123–127
148. Pitts D, Peterson K, Gilbert G et al (1996) Field assessment of irrigation system performance[J]. Appl Eng Agric 12(3):307–313
149. Branscheid VO, Hart WE (1968) Predicting field distributions of sprinkler systems [J]. Trans Am Soc Agric Eng 6:801–803
150. De Wrachien D, Medici M, Lorenzini G (2014) The great potential of micro-irrigation technology for poor-rural communities[J]. Irrig Drain Syst Eng 3(2)
151. Clark GA, Srinivas K, Rogers DH et al (2003) Measures and simulated uniformity of low drift nozzle sprinkler [J]. Trans ASAE 46(2):321–330
152. Sanchez ML, Meijide A, Garcia TL (2010) Combination of drip irrigation and organic fertilizer for mitigating emissions of nitrogen oxides in semiarid climate[J]. Agric Ecosyst Environ 137:99–107
153. Wang X, Yuan S, Zhu X (2010) Optimization of light small movable unit sprinkler system using genetic algorithms based on energy consumption indicators [J]. Trans CSAM 41(10):58–62
154. Moreno MA, Ortega JF, Corcoles JI et al (2010) Energy analysis of irrigation delivery systems; monitoring and evaluation of proposed measures for improving energy efficiency[J]. Irrig Sci 28:445–460
155. Al-Ghobari HM (2006) Effect of maintenance on the performance of sprinkler irrigation systems and irrigation water conservation [M]. Food Sci Agric Res Centre Res Bull
156. Moazed H, Bavi A, Boroomand-Nasab S, Naseri A et al (2010) Effect of climate and hydraulic parameters on water uniformity coefficient in solid set system [J]. J Appl Sci 10:1792–1796
157. Ravindra VK, Rajesh PS, Pooran SM (2008) Optimal design of pressurized irrigation subunit [J]. J Irrig Drain Eng 134(2):137–146
158. Mateos L (1998) Assessing whole field uniformity of stationary sprinkler irrigation systems [J]. Irrig Sci 18:73–81
159. Darko RO, Yuan S, Hong L et al (2016) Irrigation, a productive tool for food security- a review[J]. Acta Agric Scand Sect B-Soil Plant Sci 66(3):191–206
160. Guo RL, Xu BH (1997) Uniformity and strength of centre pivot sprinkling machine[J]. Precis Agric 7(3):21–26
161. Darko RO, Yuan SQ, Liu JP et al (2017) Overview of advances in improving uniformity and water use efficiency of sprinkler irrigation [J]. Int J Agric Biol Eng 10(2):1–15
162. Montero J, Tarjuelo JM, Carrión P (2001) SIRIAS: a simulation model for sprinkler irrigation. II. Calibration and validation of the model [J]. Irrig Sci 20:85–98
163. Li J, Kawano H (1998) Sprinkler performance as affected by the nozzle inner contraction angle [J]. Irrig Sci 18(2):63–66
164. Latif M, Ahmad F (2008) Operational analysis of water application of a sprinkler irrigation system installed on a golf course: case study[J]. J Irrig Drain Eng 134(4):446–453
165. Lan YB, Chen SD, Fritz BK (2017) Current status and future trends of precision agricultural aviation technologies [J]. Int J Agric Biol Eng 10(3):1–17
166. Tarjuelo JM; Montero J, Carrion PA (1999) Irrigation uniformity with medium size sprinklers part II influence of wind and other factors on water distribution. [J] Trans ASAE 42(3):677–689
167. Lee WS, Burks TF, Schueller JK (2002) Silage yield monitoring system. ASABE Paper No. 021165. St. Joseph, MI: ASABE
168. Wei Q, Shi Y, Dong W (2006) Study on hydraulic performance of drip emitters by computational fluid dynamics [J]. Agric Water Manag 84:130–136
169. Vories ED, Von Bernuth RD, Mickelson RH (1987) Simulating sprinkler performance in wind [J]. J Irrig Drain Eng 113(1):119–130
170. Stern J, Bresler E (1983) Nonuniform sprinkler irrigation and crop yield [J]. Irrig Sci 4:17–29
171. Baustiata C, Salvador R, Burguete J et al (2009) Comparing methodologies for the characterization of water drops emitted by an irrigation sprinkler[J]. Trans ASABE 52(5):1493–1504
172. Drocas A (2009) Determination of distribution uniformity for EEP-600 sprayer equipped with IDK 120–02 nozzle [J]. Scientific papers USAMV Bucharest, Series A. Agronomy 52:304–309

173. Dorr GJ, Hewitt AJ, Adkins SW (2013) A comparison of initial spray characteristics produced by agricultural nozzles [J]. Crop Prot 53:109–117
174. Estrada C, Gonzalez C, Aliod R et al (2009) Improved pressurized pipe network hydraulic solver for applications in irrigation systems [J]. J Irrig Drain Eng 135(4):421–430
175. Thompson AL, Gilley JR, Norman JM (1993) A sprinkler water droplet evaporation and plant canopy model 11 [J]. Trans ASAE 36(3):743–750
176. Uddin J, Hancock NH, Smith RJ et al (2013) Measurement of evapotranspiration during sprinkler irrigation using a precision energy budget (Bowen ratio, eddy covariance) methodology[J]. Agric Water Manag 116:89–100
177. Abo-Ghobar HM (1994) Effect of riser height and nozzle size on evaporation and drift losses under arid conditions [J]. J King Saud Univ Agric Sci 2:191–202
178. Huang X, Wu F, Fan Y (2006) Adjustment of elevation angle of nozzle and its effect ion hydraulics of whirl sprinkler [J]. J Drain Irrig Mach 5:29–32
179. Medici M, Lorenzini G, De Wrachien D (2013) Water droplet trajectories in a sprinkler jet flow the quantum hydrodynamic framework[J]. Irrig Drain 38(3):111–122
180. Seginer I, Nir D, Bernuth RV (1991) Simulation of wind-distorted sprinkler patterns [J]. J Irrig Drain Eng 117(2):285–306
181. Teske ME, Thistle HW, Londergan RJ (2011) Modification of droplet evaporation in the simulation of fine droplet motion using AGDISP [J]. Trans ASABE 54(2):417–421
182. Martin CE, Newman J (1991) An analytical model of water loss in sprinkler irrigation [J]. Appl Maths Comput 43:19–41
183. Molle B, Tomas S, Hendawi M (2012) Evaporation and wind drift losses during sprinkler irrigation influenced by droplet size distribution [J]. Irrig Drain 61:240–250
184. Seginer L, Kantz D, Nir D (1991) The distortion by the wind of the distribution patterns of single sprinklers [J]. Agric Water Manag 19:314–359
185. Pair CH (1968) Water distribution under sprinkler irrigation[J]. Trans ASAE 11:314–315
186. Han S, Evans RG, Kroeger MW (1994) Sprinkler distribution patterns in windy conditions [J]. Trans ASAE 37(5):1481–1489
187. Agassi M, Bloem D, Ben- M (1994) Effect of drop energy and soil and water chemistry on infiltration and erosion[J]. Water Resour 30(4):1187–1193
188. Frost KR, Schwalen HC (1955) Sprinkler evaporation losses[J]. Agric Eng 36(8):526–528
189. Assouline SM (1997) Modelling the dynamics of seal formation and its effect on infiltration as related to soil and rain fall characteristics[J]. Water Resour Res 33(7):1527–1536
190. Versteeg HK, Mallasekera W (1995) An introduction to computational fluid dynamics. In: The finite volume method [M]. New York, Wiley
191. Yuan SQ, Darko RO, Zhu XY et al (2017) Optimization of movable irrigation system and performance assessment of distribution uniformity under varying conditions [J]. Int J Agric Biol Eng 10(1):72–79
192. Tarjuelo JM, Valiente M, Lozoya J (1992) Working conditions of a sprinkler to optimize the application of water [J]. J Irrig Drain Eng 118:895–913

Chapter 2
Optimization of the Fluidic Component of Complete Fluidic Sprinkler and Testing of the New Design Sprinkler

Abstract A newly designed dynamic fluidic sprinkler was developed to improve hydraulic performance of the existing complete fluidic sprinkler under low-pressure conditions. This study presents the orthogonal test of the newly designed dynamic fluidic sprinkler with different types of nozzles at different operating pressures. The following conclusions were made: These experiments confirmed the optimal values of the dynamic fluidic sprinkler structural parameters.

Keywords Sprinkler irrigation · Orthogonal experiment · Structural parameter · Uniformity

2.1 Introduction

Sprinkler irrigation technology has been widely used especially in agriculture to save water. It has great potential for improving the water use efficiency of crops. Furthermore, the irrigation engineer can control the amount of water applied, and it is more easily scheduled, which can increase water productivity per the unit of water consumed [1, 2]. The sprinkler irrigation system distributes water in the form of discrete drops travelling through the air [3]. Sprinkler irrigation can play a significant role in irrigation development in third world countries, if the system is properly selected, designed and operated. Sprinkler systems have accelerated and been revolutionized with the development of irrigated agriculture in several parts of the world. It is therefore not surprising that the utilization of sprinkler irrigation systems has recently increased [4, 5].

According to [6, 7], the performance of a sprinkler is determined by its discharge, wetted radius, distribution pattern, application rate and droplet sizes. Water application rate can be defined as the depth of water applied to the area per unit time. It determines which sprinkler should be assigned to a particular soil, crop and terrain on which it operates. The application rate depends on the operating pressures, the nozzle size and distance between sprinklers [8]. However, the effect of operating pressure on application rate is minimal compared to the effect of the sprinkler nozzle on the application rate [3]. For most sprinklers, when the operating pressure is increased, the discharge tends to balance the increase in wetted area. It has been found that a

X. Zhu et al., *Dynamic Fluidic Sprinkler and Intelligent Sprinkler Irrigation Technologies*, Smart Agriculture 3, https://doi.org/10.1007/978-981-19-8319-1_2

sprinkler nozzle that produces little droplets covers a smaller wetted area, which also has the highest average application rate. Increasing the nozzle diameter increases the average application rate, since the sprinkler discharge tends to increase more rapidly than wetted area [9]. According to [10, 11], the application uniformity of a sprinkler is an important performance criterion for the design and evaluation of sprinklers, which is primarily influenced by operating pressure, sprinkler size and spacing.

Several studies have been conducted to analyze the droplet size distribution with different types of sprinklers over the years. The work in [12] showed that drop size distributions have a direct effect on irrigation water kinetics energy and wind drift. The work in [13] analyzed the droplet size characteristics of a complete fluidic sprinkle and concluded that about 50% of the droplets had a diameter of less 0.5 mm and that 50% of the water volume consisted of droplets with a diameter less than 2 mm at most distances. The work in [14] reported that nozzle size and pressure configurations have an influence on droplet formation. Similarly, [15, 16] reported that drop sizes can also influence the design, uniformity and efficiency of irrigation systems. According to [17], wind speed has been found to affect fine drops more than large drops. The work in [18] showed that small drops are subject to large evaporation losses under high vapor pressure. However, when drop evaporation is controlled by air friction, large drops can account for most evaporation losses [19]. The work in [20] reported that drops produced by a sprinkler are subject to several factors; such as the type of sprinkler and nozzle, operating parameters and environmental conditions.

Other researchers have proposed equations to express the coefficient of uniformity [21, 22]. The different equations available to express the coefficient of uniformity (CU) are based on some measures of variation in water distribution. The work in [23, 24] considered a coefficient of uniformity value of less than 85% as "low" and a CU of 85% or above as "desirable". According to [25, 26], Christiansen's coefficient of uniformity is the most widely used for water distribution uniformity assessment in sprinkler irrigation.

Over the years, extensive research works have been carried out to improve the structure and efficiency of the fluidic sprinkler for crop production. The work in [27] conducted experiments on drop size distributions and droplet characterization of a complete fluidic sprinkler with different nozzle dimensions. The work in [28] performed a numerical simulation and experimental study on a new type of variable-rate fluidic sprinkler. The work in [29] researched the field performance characteristics of a fluidic sprinkler. The work in [30] compared fluidic and impact sprinklers based on hydraulic performance. The work in [31] analyzed smoothed particle hydrodynamics and its applications in fluid–structure interactions. The work in [32] concluded that variations in quadrant completion times were small for both fluidic and impact sprinklers. However, deviations in water application rate were higher with the fluidic sprinkler. The work in [29] studied the relationship between rotation speed and operating pressure and pointed out that the inner angle of a fluidic sprinkler varied in quite a range among geometrical parameters. Subsequently, the authors concluded that further study needed to be carried out on the design features of the fluidic component. Similarly, Liu et al. [28] carried out a study on the fluidic sprinkler and confirmed the need to optimize the structure.

Only a few studies have focused on improving the rotation of the fluidic sprinkler. However, the rotation instability remains a major difficulty, resulting in the variation of the water application rates. Optimization can enhance the rotation stability and minimize the inconsistency in the water application rates. Therefore, the aim of this paper was to evaluate the hydraulic performance of the newly designed dynamic fluidic sprinkler with different types of nozzles at different operating pressures.

2.2 Complete Fluidic and Outside Signal Sprinklers

Figure 2.1a and b, show the schematic diagram of complete fluidic and outside signal sprinklers, respectively. In the figures, the main differences between the outside signal and a complete fluidic sprinkler are the working principle. The outside signal receives a signal when the jet flows impact on the signal device, located outside of the nozzle in the flow direction. But the complete fluidic sprinkler obtains a signal from the fluidic component, found in the inside of the working area. In previous studies, the authors made efforts to improve the performance of the fluidic sprinkler. However, the rotation instability remains a major difficulty, resulting in variations of the water application rates. Therefore, it is necessary to redesign the fluidic structure of the fluidic sprinkler by considering the contraction angle, the shape, and the size of the signal air hole. The aforementioned parameters used by previous are shown in Table 2.1.

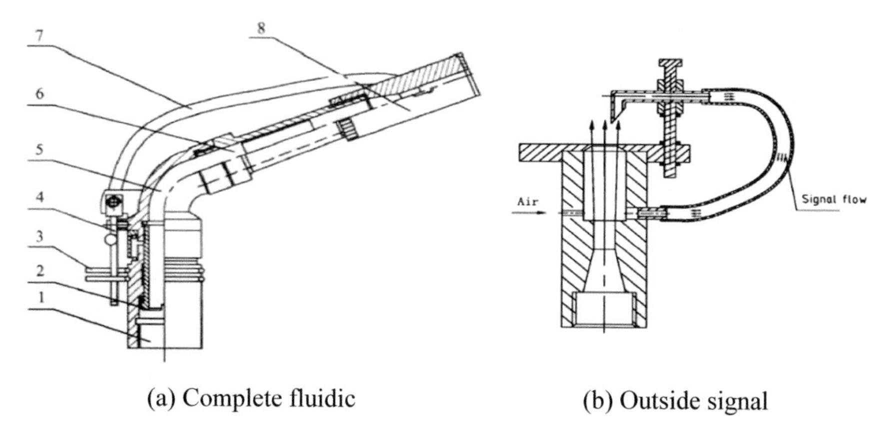

(a) Complete fluidic (b) Outside signal

Fig. 2.1 Complete fluidic and outside signal sprinklers. 1. Swivel connection block; 2. Hollow shaft; 3. Limiting ring; 4. Reverse mechanism; 5. Signals water into faucets; 6. Sprinkler tubing; 7. Reversing plastic tube; 8. Fluidic element

Table 2.1 Design parameters for outside signal and complete fluidic sprinklers

Parameter	Dimension	
	Outside signal	Complete fluidic
Contraction angle (α)	10°	20°
Diameter of base hole *(M)*	20 mm	8 mm
Offset length *(S)*	2.4 mm	2.8 mm
Working area *(L)*	43 mm	28 mm

2.3 Design of Newly Dynamic Fluidic Sprinkler Head and Working Principle

2.3.1 Working Principle

Figure 2.2 presents the structure of dynamic fluidic sprinkler. The profile of the fluidic element was defined by the inner contraction angles, the offset length, and the working area. A prototype of the dynamic fluidic sprinkler was self-designed and locally machined by using a wire-cut electric discharge machining process. The manufacturing tolerance for the size was ±0.02 mm as shown in Fig. 2.2. The working theory of (DFS) is based on the theory of the Coanda effect. The dynamic fluidic sprinkler receives an air signal from a signal tank. The working principle of the fluidic sprinkler is as follows: as water is ejected from the nozzle of the main tube into the signal thank, a region of low-pressure forms on both sides at the entry into the main jet flow. Fluid flow from the reversing plastic tubing (left) into the right side, forces the jet to deflect towards the right boundary of the signal thank where it eventually attaches. The air gap between the exit at the right side of the element and the water jet is filled by air, such that the pressures on both sides of the main jet are equal. At the same time, the nozzle receives the signal water on the left edge of the water jet, then the signal water flows in the tube to the inlet signal. Taking out the water from the contact signal into the inlet, the small gap is eventually blocked, forming a low-pressure region. When the pressure difference reaches a certain value, the main jet flows to the right-side attachment wall, Water flows from the diameter into the action zone, and the main jet is ejected from the central circular hole. Under the wall attached condition, due to the bending of the main jet, the signal water nozzle is void and no signal water is received, only air is received. After the signal water in the tube is pumped out, the air enters the water inlet through the tube. The pressures on both sides become equal. Alternate air movement from the signal nozzles and the plate cover account for the stepwise rotation [32].

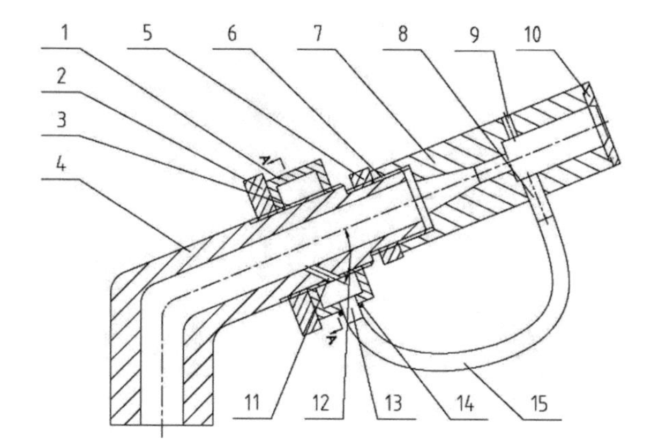

Fig. 2.2 Schematic, pictorial view of the newly fluidic sprinkler head. 1. Water signal tank. 2. First, lock nut. 3. Pipe sprayer 4. Spray body. 5. Second lock nut. 6. Body of the fluidic element. 7. Jet element body. 8. Water inlet. 9. Air hole. 10. Outlet cover plate. 11. Water dividing hole. 12. α degree. 13. Signal nozzle. 14. Third lock nut. 15. Conduit

2.3.2 Design of the Nozzles

The equipment and design factors in the sprinkler irrigation system include the nozzle characteristics which are composed of nozzle size, nozzle type, discharge angle, jet straightening vane inside the main nozzle, the number of nozzles and operating pressure. Most sprinklers have two nozzles, the main nozzle and an auxiliary nozzle that discharge water in the form of a jet into the air. Nozzles convert the pressure within the piping system into velocity upon exist from the sprinkler. The wetted coverage area and the application pattern are determined by the nozzle design and the type of sprinkler. Other researchers have studied the influence and measurement of nozzle shape on sprinkler droplet size and water application. Several types of nozzles have been developed for fluidic sprinkler including constant- diameter, and diffuse-jet. Table 2.2 presents nozzles size and corresponding pressures for the previous study, looking at the increasing cost of energy and the growing demand to saving water for optimum crop production, it is more convenient to design new nozzles size for the study. Therefore, the test nozzles were self-designed and locally machined using a wire-cut electric discharge machining (EDM) process. The prototypes of the nozzles are shown in Fig. 2.3. The inlet diameter of the nozzle was set as 15 mm, while the outlet diameters were chosen as 3, 4, 5, 6, and 7 mm.

Table 2.2 Nozzles size and corresponding pressures for the previous study

Sprinkler type	Nozzle diameter (mm)	Pressure/kPa
10PXH	4	250
15PXH	6	300
20PXH	8	350
30PXH	10	400
40PXH	14	450
50PXH	18	500

(a) DFS sprinkler head (b) Plate cover

Fig. 2.3 Prototype of the dynamic fluidic sprinkler and nozzle sizes

Table 2.3 Factors and levels

Level	Factors			
	A length of tube, *l* (mm)	*B* Pressure h, (mm)	C diameter of tube, *m* (mm)	*D* Nozzle diameter, *n* (mm)
1	15	150	2	5
2	20	200	3	6
3	25	250	4	7

2.3.3 Experimental Setup and Procedure

The hydraulic performances of structural parameters of the dynamic fluidic sprinkler were studied using an orthogonal test to determine influencing factors, the order of importance and optimal combination of the factors. The structural parameters of the sprinkler used for the study were the length of the tube (L), pressure (H), the diameter of the tube (M), nozzle diameter (N), and are represented by Factors A, B, C, and D, respectively. An orthogonal array with four factors and three levels was selected for the test as shown in Table 2.3. The tests were conducted in the sprinkler laboratory of the Research Center of Fluid Machinery Technology and Engineering, Jiangsu University in China. The laboratory is circular-shaped with a diameter of 44 m. The materials used for the experiment include; centrifugal pump, electromagnetic flow meter, and piezometer, valve and the impact sprinkler. The sprinkler was installed at a height of 2 m from the ground level with nozzle an elevation angle of 23°. The riser was at an angle of 90° to the horizontal from which the top of the catch cans was 0.9 m above the ground. Water was pumped from the reservoir through the main pipe and sprayed out from the nozzle. The working pressure was measured by a pressure gauge at the base of the sprinkler had an accuracy of 0.4%. The sprinkler was run for 30 min to standardize the environment conditions before the experiment

Table 2.4 Scheme used in the orthogonal test

Test	*A*	*B*	*C*	*D*
1	1	1	1	1
2	1	2	2	2
3	1	3	3	3
4	2	1	2	3
5	2	2	3	1
6	2	3	1	2
7	3	1	3	2
8	3	2	1	3
9	3	3	2	1

was carried out. Different inlet pressures were tested during the investigation and these include 150, 200 and 250 kPa. The corresponding flow rate was 1.47, 1.57 and 1.66 m^3/h for 150, 200 and 250, respectively. ASAE approach was used to determine the application of water depth measurements. The catch cans were used to collect water in an hourly base and measured with a graduated cylinder (Table 2.4).

The discharge coefficients of each nozzle were determined for the observed pressure-discharge data using Eq. (1.1).

Matrix Laboratory (MATLAB) program was used to compute the combined CU values according to the radial water distribution. Radial data of water distribution from the fixed water dispersion devices were modified into net data. The final calculated average radial water distribution data was the same in all directions from the A1, B1, C1, A2, B2, and C2. The available data points were distributed like a spider web. A grid of data points was converted to calculate the combined CU. The depth of the net point depends on the distance away from the sprinkler. The water depth of every interpolating point, assumed to be a continuous variable value, were calculated using a mathematical model of interpolating cubic splines. The uniformity of water application rate was evaluated using the Christiansen coefficient of uniformity (CU) in Eq. (1.5).

The direct analysis technique was used to analyze our test results. This technique can identify influencing factors in decreasing order of importance, and the optimal combination of factors can be forecasted. The calculation formula is as follow:

Y1X (Y2X, Y3Z) = Sum of corresponding data 1(2, 3) for

$$\text{Column}X\ j(\text{X} = \text{A, B, C, D};\ j = 1, 2, 3) \tag{2.1}$$

$$\text{YIX (Y2X, Y3X)} = \text{Average of Y1X (Y2X, Y3X)} \tag{2.2}$$

$$\text{RX} = \text{Maximal YJX minus the minimal YJX} \tag{2.3}$$

Table 2.5 Discharge coefficient for different types of nozzles

		Discharge			
Nozzle size (mm)	Pressure (kPa)	150	200	250	Standard deviation
5		0.75	0.77	0.79	0.0163
6		0.85	0.76	0.66	0.077
7		0.70	0.56	0.59	0.060

2.3.4 *Results and Analysis of Orthogonal Tests*

2.3.4.1 Comparison of Operating Pressure and Discharge

Table 2.5 presents the results of measured flow rates of sprinkler irrigation nozzles used in this study under different operating pressures. Analysis of the measured data was performed to find the influence of the geometrical parameters as well as the operating pressure on the discharge of the sprinkler. As shown in Table 2.5, when using the sprinkler, the measured nozzle flow rates ranged from 1.4to 1.47 m^3/h with a mean value of 1.435, 1.5 to 1.57 m^3/h with a mean value of 1.535, and 1.59 to 1.66 m^3/h with a mean value of 1.625, for 150, 200 and 250 kPa, respectively. The coefficient of discharge for 5 mm nozzle ranged from 0.75 ~ 0.79 with an average value of 0.77, while that from the 6 and 7 mm was from 0.66 ~ 0.85, 0.56 ~ 0.59 with an average of 0.756 and 0.62, respectively. From the analysis, it was established that the coefficients of discharge fluctuated within a small acceptable range under the same operating pressures. The coefficients of discharge obtained using 4 mm nozzle were higher than those obtained using the 5 and 6 mm nozzles, which means that the 4 mm nozzle had the advantages of higher irrigation intensities. These can be attributed to fewer restrictions within the inner flow movement. It can be confirmed that the discharge coefficient does not depend on the operating pressure. Similar results were published by.

2.3.4.2 Summary Results of the Orthogonal Test

Tables 2.6 and 2.7 and present the results of factors influencing the CUs and the spray range. From the study, the relatively ideal results should be higher uniformity coefficient and spray range. It was revealed that test 2 (A1 B1 C1 DI), test 6 (A2 B3 C1 D2), and test 7 (A3 B1 C3 D2), had the highest uniformity coefficient. This could be attributed to the fact that flow rate at same pressure was much higher and turbulence flow was less uniform resulting in better distribution. Test 1 (A1 B1 C1 DD1), test 2 (A1 A2 C2 A2), and test 7 (A3 B1 C3 D2) also had the highest spray range. Test 3 (A1 B3 C3 D3), test 4 (A2 B1 C2 D3) and test (A2 B2 C3 D1) were normal. Test 8 (A3 B2 C1 D3), and test 9 (A3 B3 C2 D1) were not effective because low uniformity and spray range were observed. As shown in Table 2.6, a higher *R*-value shows that the factor had a strong effect on the test results, which means that

Table 2.6 Test scheme and results

Test number	A	B	C	D	CU (%)	Range(m)
1	1	1	1	1	85	12
2	1	2	2	2	86.5	12
3	1	3	3	3	85.5	8
4	2	1	2	3	85.5	8
5	2	2	3	1	83.9	11
6	2	3	1	2	86	8
7	3	1	3	2	91	13
8	3	2	1	3	85	7
9	3	3	2	1	83.4	9

Table 2.7 Results of structural parameter combination

		A	B	C	D
CU	Yj1	240.00	242.50	241.20	236.35
	Yj2	239.60	239.40	240.45	244.70
	Yj3	240.45	140.15	240.40	241.00
	Yj1	80.67	80.83	80.40	78.78
	Yj2	79.87	79.80	80.15	81.57
	Yj3	80.15	80.05	80.13	80.33
	R	0.80	1.03	0.27	2.78
Range	Yj1	31.00	31.00	27.00	32.00
	Yj2	25.00	29.00	26.00	32.00
	Yj3	29.00	35.00	22.00	32.00
	Yj1	10.30	10.30	9.00	10.67
	Yj2	8.33	9.67	8.67	10.67
	Yj3	9.67	10.33	10.67	7.00
	R	2.00	2.00	2.00	3.67

the factor is significant, a lower *R*-value indicates that the factor had a weak effect on the test results, which means that it is not significant. The following observations can be drawn from the results shown in Table 2.6 and Fig. 2.4.

Factor A: When the length of the tube was varied from 20 to 25 mm, the CUs varied from 85% to 86.5% with an average value of 85.6 (A = 15 mm), from 83.9% to 87% with an average of 85.3 (A = 20), and from 83.4% to 91% with an average of 86.46% (A = 25 mm). The range also varied from 8 to 12 m with an average of 10.3 m (A = 15 mm), from 6 mm to12mm with an average of 8.3 m (A = 20 mm), and from 7 to 13 mm with an average of 9.6 m (A = 25 mm). The sprinkler worked perfectively at a length of 25 mm, but when the nozzle was 6 mm regardless of

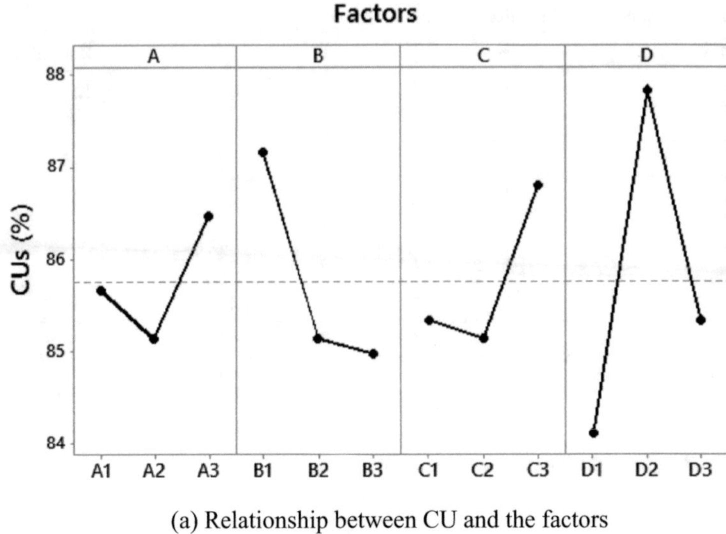

(a) Relationship between CU and the factors

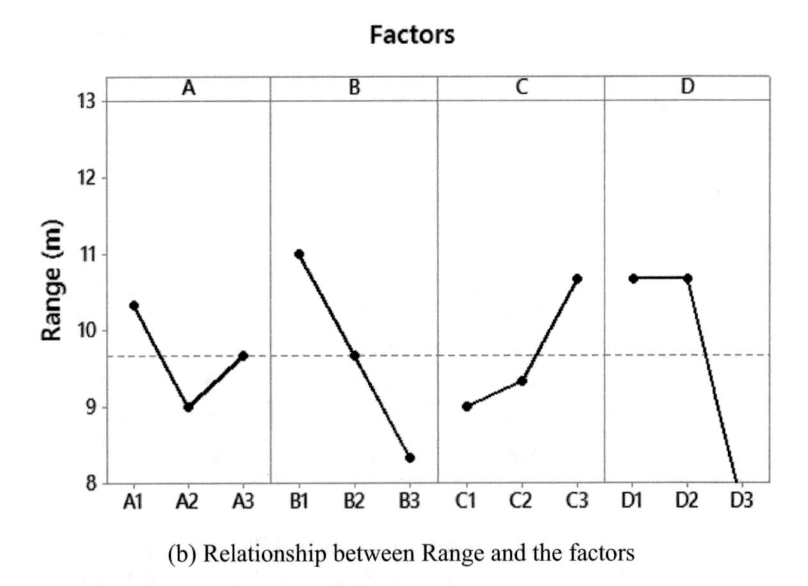

(b) Relationship between Range and the factors

Fig. 2.4 Factors influencing CUs and range

structure parameter changes, the sprinkler was not effective when the length of the tube was less than 20 mm or greater than 25 mm.

Factor B: When operating pressure was varied from150 to 250 kPa, the CUs increased from 85 to 91% with an average of 87.16% (B = 150 kPa), from 85.5% to 86.5% with an average of 85.7% (B = 200 kPa), and from 83.2% to 86.2% with an average of 84.9% (B = 250 kPa). The spray range varied from 6 to 13 m with an average of 10.3 m (B = 150 kPa), from 7 to 11 m with an average of 8 m (B = 200 kPa), and from 8 to 9 m with an average value of 8.33 m (B = 250 kPa). The CUs and spray were significantly reduced with increasing working pressure. These can be attributed to restriction of the flow of water within the sprinkler resulting in a low amount of water application.

Factor C: When the diameter of the tube was varied from 2 to 4 mm, the CUs changed from 85 to 87% with an average of 85.66% (C = 2 mm), from 83.4% 86.5% with an average of 85.0% (C = 3 mm), and from 83.91% to 91% with an average of 86.8% (C = 4 mm). The spray range varied from 7 to 12 m with an average of 9 m (C = 2 mm), from 6 m to11m with an average of 8.67 m (C = 3), and from 8 to 13 m with an average of 10.7 m (C = 4 mm). The CU was highest at 3 mm because the overlaps were higher and the water distribution was more uniform when the tube was narrow.

Factor D: The nozzle diameter was varied from 4 to 6 mm. The CUs changed from 83.9% to 85% with an average of 84.1% (D = 4 mm), from 86.5% to 91% with an average value of 88.16% (D = 5 mm), and from 80% to 80.5% with an average value of (D = 6 mm). The range varied from 9 to 12 m with an average of 10.66 m (D = 4 mm), from 11 to 13 m with an average of 10.7 m (D = 5 mm), and from 6 to 8 m with an average of 7 m (D = 6 mm). The CUs and spray range decreased as the diameter of the nozzle was increased because a larger part of the jet flow was uninterrupted particularly in the case of the nozzles with the diameters of 6 mm. The optimal nozzle diameter was found to be 5 mm. The comparison of the test scheme indicated that CUs exceeded 85% in 5 tests, and the range exceeded 10 m in 4 tests. Tests 7 and 2 were ideal. Test 7 (A1B1C1D2) had the highest uniformity coefficient and the longest range. Test 5 was not effective because B was too small when D was a lager. Test 9 was also ineffective because B was too large when D was small. The optimal combination of structural parameters was achieved with the factor combination of A3B1C3D1.

2.3.4.3 Simulation of Water Distribution

Figure 2.5 presents the plots of water distribution for all the 9 tests. It can be observed that variations in the contour and color maps around the sprinkler had different application rates. Rings that have similar color indicate uniform water distribution pattern, whiles different colors in the ring represent non-uniform water distribution patterns.

Comparison of water distribution for the various tests showed that test 7 and test 2 produced a high uniformity for a given operating pressure which corresponds to the orthogonal results. However, test 7 was slightly higher compared to test 2. It is possible that after interruption with the alignment signal nozzle the flow became less uniform, leaving more water applied near the sprinkler. This means that test 7 can improve the uneven distribution of water and save water for crop production. As a consequence, differences in water distribution can be seen in most areas around the sprinkler in the case of the other tests, which is in agreement with [33] reported

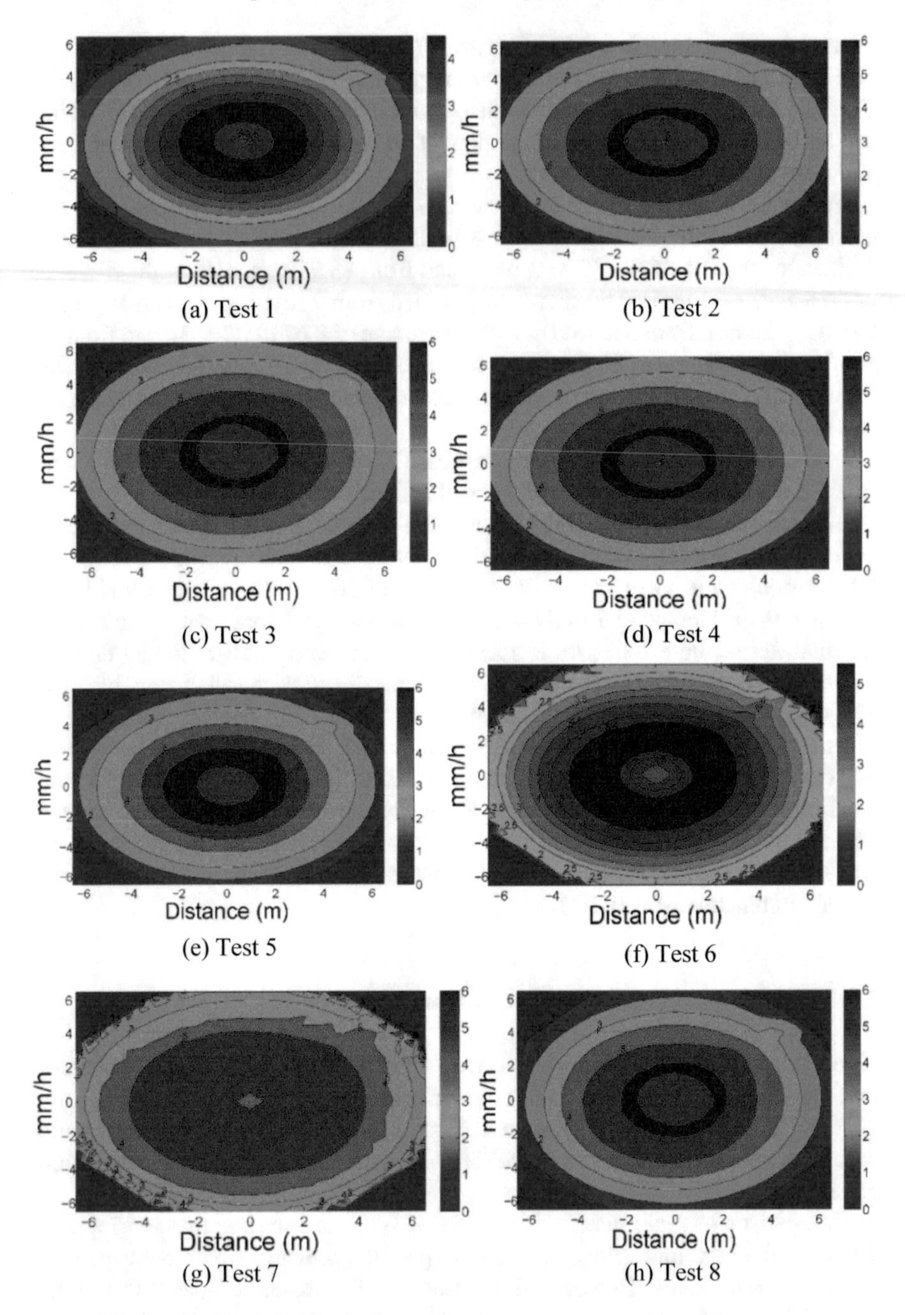

Fig. 2.5 An illustrative example of water distribution maps for tests 1 through 9

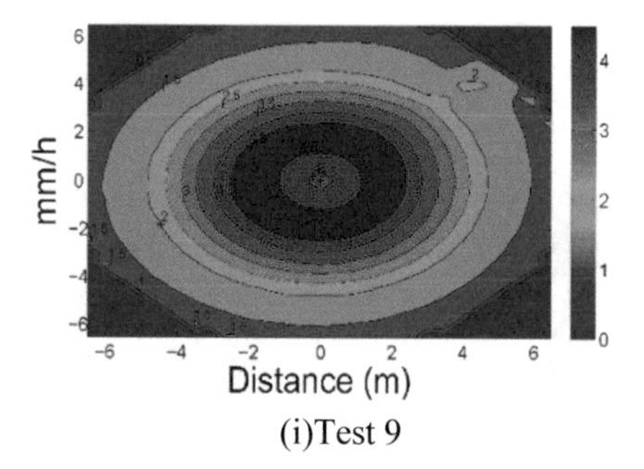

Fig. 2.5 (continued)

similar results in their experiments using fluidic sprinkler results of variations in rotations speed with respect to the quadrants.

2.3.5 Brief Summary

This study presents the orthogonal test of the newly designed dynamic fluidic sprinkler with different types of nozzles at different operating pressures. The following conclusions were made: These experiments confirmed the optimal values of the dynamic fluidic sprinkler structural parameters. The length of the tube is 25 mm, the working pressure is 150 kPa, the diameter of the length of the tube is 3 mm, and the nozzle diameter is 5 mm. The factors influencing the CU and range in decreasing order of importance were nozzle diameter, pressure, length of the tube and the diameter of the tube. The optimal combination of structural parameters was achieved with the factor combination of A3B1C3D1.

2.4 Evaluation of Hydraulic Performance Characteristics of a Newly Designed Dynamic Fluidic Sprinkler

2.4.1 Design of New Dynamic Fluidic Sprinkler Head

In this research, a newly designed dynamic fluidic sprinkler head was manufactured. The following parameters are key factors when it comes to the design of the fluidic structure: the diameter of the main nozzle, the inner contraction angle, the offset length and the working area. The dynamic fluidic sprinkler was developed by Jiangsu University. It is schematically shown in Fig. 2.6. The manufacturing tolerance for the size was ± 0.02 mm. The main differences between the newly designed dynamic fluidic sprinkler (DFS) and complete fluidic sprinkler (CFS) is the working principle.

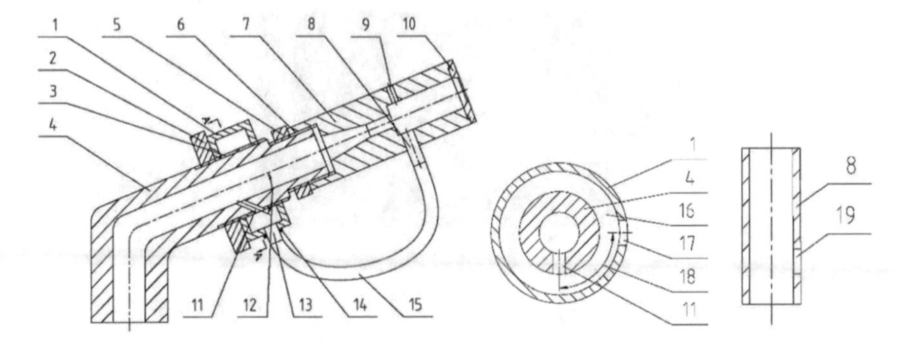

Fig. 2.6 Schematic, pictorial view of the new fluidic sprinkler head. 1. Water signal tank. 2. First lock nut. 3. Pipe sprayer 4. Spray body. 5. Second lock nut. 6. Body of the fluidic element. 7. Jet element body. 8. Water inlet. 9. First air hole. 10. Outlet cover plate. 11. Water dividing hole. 12. α main flow.13. Signal nozzle. 14. Third lock nut. 15. Conduit. 16. Water storage capacity. 17. Signal hole. 18. β contraction angle. 19. Second air hole

The newly designed dynamic fluidic sprinkler receives an air signal from a signal tank, but the complete fluidic sprinkler obtains the signal from the fluidic component, found in the working area. When they are operating under a low pressure condition (such as 100 kPa), it is difficult to get the signal flow for the complete fluidic sprinkler. This leads to disappearance of the pressure difference between the two sides of the wall. Therefore, the CFS rotation could not be guaranteed. For the DFS, the air signal flow could be received continuously once the signal tank is filled with water.

2.4.2 Working Principle

The principle of operation of the fluidic sprinkler is based on [34] to perform the function of rotation. Water is ejected from the main nozzle to the working area. A region of low-pressure eddy is formed on both sides of the working area. Air flows into the left side from the reverse blow down nozzle and into the right side from the signal nozzle. The main flow jet is straight because the pressure on both sides is equal and the sprinkler remains stationary, as shown in Fig. 2.7a,b, respectively. The signal flow received from Signal Nozzle 1 fills up Signal Nozzle 2 to transform the right side into a low-pressure eddy. The main flow jet is bent toward the boundary and eventually attached to it because the left pressure is much larger than the right pressure. The phenomenon is repeated step by step, and the sprinkler achieves a stepwise rotation in sequence by self-control. The main flow jet is reattached to the left plane, and the sprinkler rotates to the opposite direction because the right pressure is much larger than the left pressure. The reverse blow down nozzle opens, and air flows into the left side to equalize the pressure again when the sprinkler rotates to the other side.

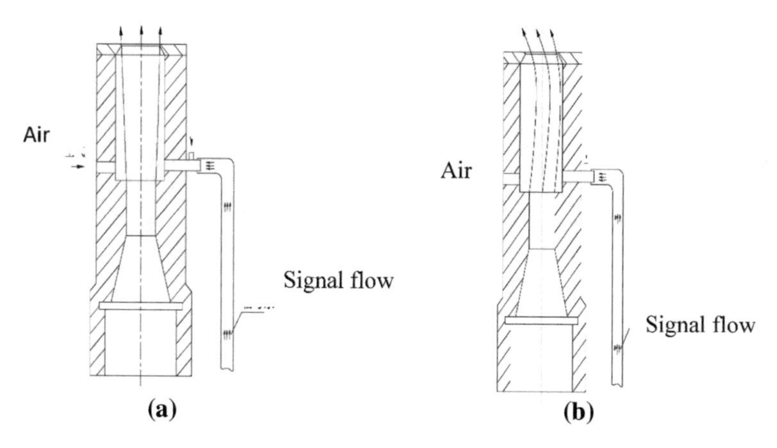

Fig. 2.7 **a** Straight main flow jet. **b** Main flow jet reattached to the right

2.4.3 Experimental Procedures

The experiments were conducted at the indoor facilities of the Research Center of Fluid Machinery Engineering and Technology, Jiangsu University (Jiangsu province). The diameter of the circular-shaped indoor laboratory was 44 m. A centrifugal pump was used to supply water from a constant level reservoir. The sprinkler head was mounted on a 1.5 m riser at a 90° angle to the horizontal. Catch cans used in performing the experiments were cylindrical in shape, 200 mm in diameter and 600 mm in height. The catch cans were arranged in two legs around the sprinkler as shown in Fig. 2.8. Each leg contained 14 catch cans placed 1 m apart constituting 28 catch cans in total. The sprinkler was run for some minutes to standardize the environment conditions before the experiment was carried out. The sprinkler flow rate was 4.75 m^3/h for an operating pressure of 250 kPa, which was controlled by pressure regulation. The operating pressure at the base of the sprinkle head was regulated and maintained by a valve with the aid of a pressure gauge with an accuracy of $\pm$ 1%.The corresponding operating pressures were 100, 150, 200, 250 and 300 kPa, respectively. The application of water depth measurements was carried out in accordance with [31].

The experiment lasted for an hour, and the water depth in the catch cans was measured with a graduated measuring cylinder. Droplet sizes were determined using a Thies Clima Laser Precipitation Monitor (TCLPM). It has the following specification: the drop diameter measurement ranges from 0.125 to 8.5 mm in increments of 0.125 mm, and the measuring area is 228 mm long, 200 mm wide with a thickness of 0.75 mm, manufactured by Adolf Thies GMBH & CO. KG, Gottingen, Germany. The principle of operation is such that a beam of light is produced from a laser-optical source in the form of infrared, 785 nm. A photo-diode with a lens is located on the receiver side to determine the optical intensity after transformation into electrical signals. The receiving signal reduces when the water droplet falls through the

Fig. 2.8 Experimental setup in the indoor laboratory

measuring area. The diameter of the droplet is estimated from the amplitude of the reduction, the droplet velocity of which is calculated from the duration of the reduced signal. For each operating pressure, the droplet size distributions were determined at an interval of 2 m along a radial transect at a distance of 2 m from the sprinkler. For each droplet measurement, the sprinkler was allowed to rotate over the TCLPM at least five times to ensure a sufficient number of drops passed through the measured area. At each pressure, a minimum of three replication assessments were made, and the averaged data were used for the final experiments. Data were ordered according to the drop diameter.

2.4.4 Computed Coefficient of Uniformity

Matrix Laboratory (MATLAB R2014a) software manufactured by Mathwork Incoperation, Springfield, MA, USA was employed to establish a computational program for the CU. The work in [25, 26] reported that Christiansen's coefficient of uniformity is the most widely used and accepted uniformity criterion. Therefore Christiansen's equation was utilized to determine CU.

$$CU = \left(1 - \frac{\sum_{i=1}^{n} |X_i - \mu|}{\sum_{i=1}^{n} X_i}\right) 100\% \tag{2.4}$$

where n = number of catch cans; x_i = measured application depth, mm; μ = mean average depth, mm; and CU = coefficient off uniformity, %.

The model for converting radial data into the net data's insert function was established as follows: Point A is the net point between two adjacent radial rays, and (X_k, Y_k) is its coordinate. P_1, P_2, P_3 and P_4 are the four nearest points to Point A on the adjacent radial rays, and $(P_1 Q_1)$, $(P_2 Q_2)$, $(P_3 Q_3)$ and $(P_4 Q_4)$ are their coordinates. Their positions are therefore $x_1 = P_1 \cos\varnothing_1$, , $(i = 1, 2, 3, 4)$ and $y_1 = P_1 \sin\varnothing_1$, $(i = 1, 2, 3, 4)$; their water depths are h_1, h_2, h_3 and h_4; and the distances away from Point A are r_1, r_2, r_3 and r_4, respectively. Thus,

$$r_1 = \sqrt{(Xi - Xk)^2 + (Yi - Yk)^2}(i = 1, 2, 3, 4) \tag{2.5}$$

The water depth A can be expressed as:

$$h_A = C_1 h_1 + C_2 h_2 + C_3 h_3 + C_4 h_4 \tag{2.6}$$

where:

$$C_1 = (r_2 r_3 r_4)^2/R,\ C_2 = (r_1 r_3 r_4)^2/R,\ C_3 = (r_1 r_2 r_4)^2/R,\ C_4 = (r_1 r_2 r_3)^2/R, \tag{2.7}$$

$$R = (r_1 r_2 r_3)^2 + (r_1 r_3 r_4)^2 + (r_1 r_2 r_4)^2 + (r_2 r_3 r_4)^2 \tag{2.8}$$

According to the actual measurements, the water depth of every point can be calculated using Eq. (2.5). The combined coefficient of uniformity can then be calculated for the overlapping of the spray sprinkler with different lateral spacings.

Basic drop statistics: Managing the large dataset obtained from the photographs required a statistical approach. While it is convenient to represent the sets by a reduced number of parameters, some traits of the drop populations can be obscured by the choice of statistical parameters. The parameters used in this work for drop diameter included arithmetic mean diameter (Eq. (2.9)), standard deviation (Eq. (2.11)) and coefficient of variation (Eq. (2.12)). The following addition parameters were determined for drop diameter: the volumetric mean (D_v) and average volumetric diameter (D_{50}).

$$\bar{d} = \frac{\sum_{i=1}^{n} m_i d_i}{\sum_{i=1}^{n} m_i} \tag{2.9}$$

$$d_v = \frac{\sum_{i=1}^{n} d_i^4}{\sum_{i=1}^{n} d_i^3} \tag{2.10}$$

$$SD_D = \sqrt{\frac{1}{n-1} \sum_{i=1}^{n} \left(d_i - \bar{d}\right)^2} \tag{2.11}$$

$$CV_D = \left(\frac{SD_D}{\bar{d}}\right) \times 100 \tag{2.12}$$

where d_i = the diameter of the droplet in each set (mm), n_i = the droplet number, i = the number of droplets in the set, $\bar{d}$ = the arithmetic mean droplet, d_v = the volume weighted average droplet diameter, SD_D = the standard deviation and CV_D = the coefficient of variation.

In order to test the difference between the means of the independent samples of 150 and 250 kPa, the study employed an independent sample *t*-test where variances were assumed to be equal with the *t*-test statistics formulated as:

$$t = \frac{(\overline{X}_1 - \overline{X}_2) - (\mu_1 - \mu_2)}{\sqrt{\frac{s_1^2}{n_1} + \frac{s_2^2}{n2}}} \tag{2.13}$$

where $\overline{x_1}$ and $\overline{x_2}$ are sample means, μ_1, μ_2 are population means, s_1^2 and s_2^2 are variances and n_1 and n_2 are the sample sizes for 150 and 250 kPa, respectively.

The above tests were carried out according to the standards of [35].

2.4.5 Results and Discussion

As shown in Table 2.8, the smallest radius of throw was obtained when the sprinkler was operated at the pressure of 100 kPa, and the maximum radius of throw was also obtained at 250 kPa for five of the six nozzles sizes tested in the present experiment. The difference between the maximum and the minimum radius of throw was 7.2 m. For all the nozzle sizes, the distance of throw increased with an increase in operating pressure until it reached 250 kPa, when it began to decrease. The distance of throw increased when the diameters of the nozzle sizes were increased, and it began to decrease for all the nozzle sizes. Similar findings were reported by [36]. This is possible because at a high pressure condition, the jet breaks up quickly, resulting in smaller radius of throw. For smaller diameters, the jet flow was restricted, resulting in a smaller radius of throw. The result from the independent sample *t*-test analysis (Table 2.9) showed that there was no significant different between radius of throw for 250 and 150 kPa since ($p > 0.05$). The obtained results for the radius of throw were similar to previous findings by Zhu et al. [32].

Table 2.8 Radius of throw for different types of nozzles and pressures

Nozzle size (mm)	Radius of throw (m)					Standard deviation (m)				
	p									
	100	150	200	250	300	100	150	200	250	300
2	6.4	7.4	7.9	8.7	8.1	0.2	0.3	0.5	0.3	0.7
3	8.5	9.7	10.7	11.7	10.7	1.2	0.1	0.6	0.2	0.4
4	11.3	12.4	13.1	12.8	11.5	0.2	0.1	0.2	0.1	0.2
5	10.3	13.3	13.5	13.6	12.5	0.3	0.2	0.1	0.1	0.2
6	6.4	6.9	7.5	8.2	7.2	0.1	0.4	0.2	0.3	0.1
7	5.3	6.3	7.4	8.4	7.5	0.1	0.2	0.4	0.5	0.4

2.4.5.1 Comparison of Water Distribution Profiles

Figure 2.9 shows the application rate profiles of the newly designed dynamic fluidic sprinkler with different types of nozzles at 100, 150, 200 250 and 300 kPa, respectively. Generally, the application rates increased with an increase in nozzle diameters for all operating pressures, and these results are in agreement with [30]. As the distance from the sprinkler increased, the application rate also increased until it got to the maximum value and decreased for all the pressures. As operating pressure was increased, the application rates increased until they reached the maximum, when they started to decrease. The application rate of the 5.5-mm nozzle varied from 5.24 to 7.42 mm h − 1. The maximum value of the application rate was obtained for the five analyzed pressures (7.6 mm h − 1 at distances of 8 m for 100 kPa, 6.1 mm h − 1 at 10 m for 150 kPa, 6.23 mm h − 1 at 7 m for 200 kPa, 6.53 mm h − 1 at 7 m for 250 kPa and 7.42 mm h − 1 at 7 m for 300 kPa). Among the pressures, 200 kPa performed slight better than 150 kPa. The result from independent sample t-test analysis indicated that there was no significant difference between 250 and 150 kPa ($p > 0.05$). The comparison of the water distribution profiles at different operating pressures showed that all the different nozzle sizes produced parabola-shaped profiles, but the 5.5-mm nozzle size was flatter than the other nozzle sizes at a low pressure of 150 kPa. This could be attributed to the fact that flow rate at the same operating pressures was much higher and the internal turbulent flow was less uniform from the nozzle outlet, as well as more water was applied near the sprinkler, resulting in a more uniform water distribution for the 5.5-mm nozzle compared to the others. Several studies have shown that [37, 38] a doughnut-shaped water distribution leads to surface runoff because more water is deposited away from the sprinkler, affecting the quality of sprinkler irrigation. This implies that a 5.5-mm nozzle size can improve the non-uniform water distribution and save water for crop production. These results are better than those obtained by earlier researchers who used a similar sprinkler type.

Table 2.9 Independent sample *t*-test

		Levene's test for equality of variances		*t*-Test for equality of means						
		F	Sig	t	df	Sig. (2-tailed)	Mean difference	Std. error difference	95% Confidence interval of the difference	
									Lower	Upper
Distance	Equal variances assumed	0.727	0.442	−2.530	4	0.065	−0.26667	0.10541	−0.55933	0.02600
	Equal variances not assumed			−2.530	3.448	0.075	−0.26667	0.10541	−0.57876	0.04542

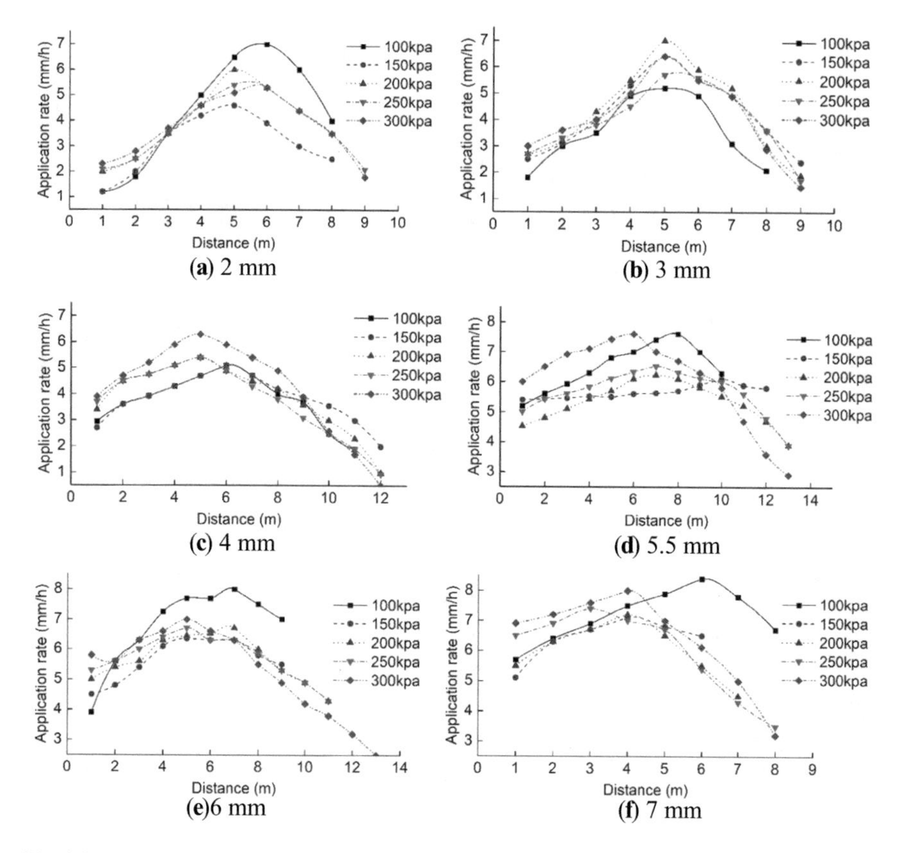

Fig. 2.9 Water distribution profiles for different types of nozzles and pressures

2.4.5.2 Comparison of the Computed Uniformity Coefficient

Figure 2.10 presents the computed coefficients of uniformity with different types of nozzles at 100, 150, 200, 250 and 300 kPa, respectively. The computed coefficients of uniformity were determined using Eq. (1). The rectangular spacing for lateral radius times of 1.0, 1.1, 1.2, 1.3, 1.4, 1.5, 1.6, 1.7, 1.8, 1.9 and 2.0 was used for all the nozzle sizes in the study. Figure 2.6 shows the relationships between the simulated CU and spacing along the vertical and horizontal axis. As the distance from the sprinkler increased, the coefficient of uniformity also increased until it got to the maximum and then decreased for all the pressures and nozzles. The average of the computed values for the 5.5-mm nozzle size was (at different pressures) as follows; 76, 81, 77, 82 and 77% and 100, 150, 200, 250 and 300 kPa, respectively. Comparatively, 250 kPa performed slightly better than 150 kPa, but 150 kPa was selected as the optimum operating pressure because of rising energy costs. For all the nozzle sizes. 5.5 mm gave the highest computed uniformity value of 86%, at a low pressure of 150 kPa. This indicates that 5.5 mm produced a better water distribution pattern

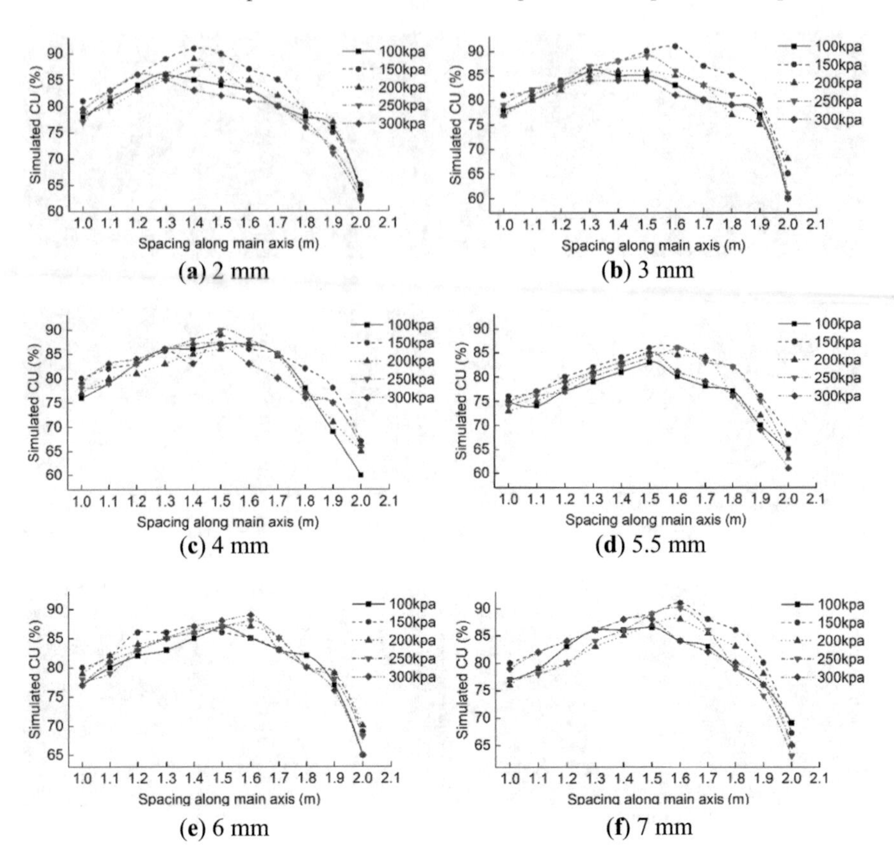

Fig. 2.10 Computed coefficient of uniformity (CU) for different types of nozzles and pressures

than the rest of the nozzles. These results are slightly better than those obtained by previous researchers for the complete fluidic sprinkler and the outside signals of 82 and 80.88%, respectively [27]. Although 250 kPa gave higher CU than 150, there was no significant difference ($p > 0.05$), along with the increasing cost of energy and growing demand for saving water for optimum crop production. It is appropriate to use 150 kPa.

The range of computed CU values for the 5.5-mm nozzle size at 150 kPa was as follows: 77% at a spacing of 1- to 68% at 2.0-times (150 kPa). The highest CU occurred at 1.6-times spacing uniformity and increased with a spacing of one- to 1.6-times, ranging from 76% to 86 with an average of 80%; subsequently, the uniformity decreased with spacing from 1.6- to 2.0-times; the CU value ranged from 84 to 68% with an average of 79.2% at an operating pressure of 150 kPa.

In general, CU values resulting from the 5.5-mm nozzle size were higher compared to other nozzles. The explanation could be that the internal turbulent flow was less uniform from the nozzle outlet and more water was applied near the sprinkler, resulting in a higher combined CU. This supports already established results from earlier research works [24, 32, 39]. The performance of the tested sprinkler was better than earlier research for the different types of fluidic sprinklers.

2.4.5.3 Droplet Size Distributions

Figure 2.11 shows the cumulative droplet diameter frequency for different types of nozzles at different operating pressures. Low operating pressures resulted in larger droplet diameters, and as operating pressures increased, smaller droplets diameters were produced. Droplet diameter increased with distance from the sprinkler for the various nozzle sizes, which is similar to previous results obtained [40].

As can be seen in Fig. 2.11, 5.5 mm gave better results than the rest of the nozzles. The average droplet diameters ranged from 0 to 3.2 mm. The cumulative frequencies were under 1 mm of 87, 67, 86.73 and 99%, under 2 mm of 89, 77, 65, 67 and

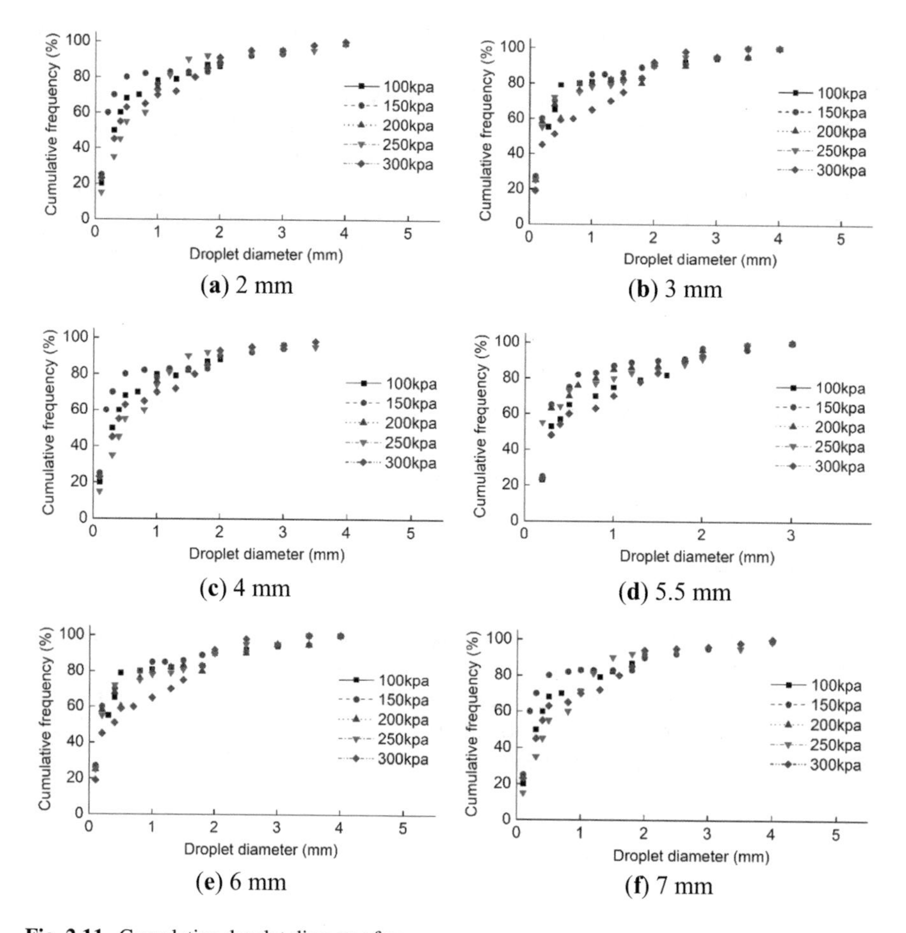

Fig. 2.11 Cumulative droplet diameter frequency

100% under 3 mm of 88, 90, 67, 88 and 55 at pressures of 100, 150, 200, 250 and 300 kPa, respectively. The mean droplet diameters for the nozzle sizes of 2, 3, 4, 5.5, 6 and 7 mm ranged from 0 to 4.2, 0 to 3.7, 0 to 3.6, 0 to 3.2,0 to 0.5 and 0 to 3.8 mm, respectively The comparison of droplet size distributions showed that 5.5 mm had the narrowest droplet size and smallest maximum droplet diameter of 3.2 mm. The biggest droplet size ranged with the maximum value of 4.2 for a nozzle size of 2 mm. These results are similar to those obtained by previous researchers who used different sprinkler types. It can also be noted that at most distances from the sprinkler, the number of droplets at smaller diameters was greater compared to that at larger diameters. This goes to support the hypothesis that droplet formation is a continuous process along the jet trajectory [41–43]. Using a 5.5-mm nozzle size will produce optimum droplet sizes, which can fight wind drift and evaporation losses. This is because large droplets possess high kinetic energy, and on impact, they disrupt the soil surface, especially soils with crustiness problems, leading to sealing of the soil surface. Dwomoah et al. reported similar results when analyzing drop diameter measurements performed with the Thies Clima Laser Precipitation Monitor (TCLPM).

Table 2.10 shows the percentage of droplets with a mean diameter for the various nozzle sizes at different operating pressures. Diameters d_{10}, d_{25}, d_{50}, d_{75} and d_{90} represent the diameters corresponding to 10, 25, 50, 75 and 90%, respectively, of the volume of detected water. From the table, it can be observed that for all the nozzle sizes, droplet size increased with increasing percentage of droplet diameter. In this experiment, almost 20% of the drops identified at all the distances from the sprinkler were smaller than the minimum diameter obtained from earlier researchers who used similar sprinkler types.

2.4.5.4 Droplet Characterization Statistics

Table 2.11 presents statistical parameters for the droplets for different types of nozzle at different operating pressures. Parameters include the arithmetic mean diameter, the volumetric mean diameter, the median diameter, the standard deviation and the coefficient of variation. All the parameters decreased with an increase in operating pressure for the nozzle sizes. All the parameters increased as the nozzle sizes increased for all the operating pressures. The mean droplet diameter and volumetric median diameter decreased with operating pressures for the nozzle sizes. Among the nozzles, 5.5 mm performed better than the rest of the nozzles. The standard deviation of the droplet diameter ranged from 0.69 to 0.86 with a mean of 0.775, and the coefficient of variation ranged from 91 to 147% with a mean value of 119% [33].

Table 2.10 Droplet sizes (mm) for 10, 25, 50, 75 and 90% ($d_{10}, d_{25}, d_{50}, d_{75}$ and d_{90}, respectively) for different types of nozzle

Nozzles size	Pressure (kPa)	d_{10}	d_{25}	d_{50}	d_{75}	d_{90}	Standard deviation (m)
2 mm	100	0.07	0.18	0.45	0.46	1.94	0.76
	150	0.05	0.14	0.36	1.09	1.55	0.65
	200	0.08	0.15	0.35	1.08	1.56	0.65
	250	0.07	0.16	0.27	1.09	1.85	0.7
	300	0.09	0.15	0.25	1.3	1.87	0.87
3 mm	100	0.06	0.13	0.36	0.47	2.09	0.83
	150	0.07	0.14	0.27	0.79	2.4	0.91
	200	0.06	0.15	0.25	0.82	2.3	0.93
	250	0.06	0.16	0.27	0.5	2.05	0.82
	300	0.09	0.18	0.25	1.49	1.96	0.87
4 mm	100	0.08	0.13	0.27	0.4	1.69	0.65
	150	0.07	0.15	0.26	0.71	1.88	0.75
	200	0.07	0.15	0.26	0.73	1.86	0.74
	250	0.08	0.17	0.26	1.02	1.82	0.74
	300	0.09	0.6	0.25	1.18	1.7	0.65
5.5 mm	100	0.04	0.11	0.34	0.44	2.05	0.69
	150	0.04	0.12	0.24	0.77	2.1	0.76
	200	0.04	0.12	0.23	0.79	2.1	0.75
	250	0.04	0.13	0.23	0.48	2.02	0.76
	300	0.05	0.14	0.23	0.47	1.93	0.70
6 mm	100	0.04	0.11	0.34	0.44	2.05	0.83
	150	0.05	0.12	0.24	0.77	2.1	0.85
	200	0.05	0.13	0.23	0.79	2.1	0.85
	250	0.05	0.14	0.23	0.48	2.02	0.81
	300	0.07	0.14	0.23	1.47	1.93	0.87
7 mm	100	0.05	0.16	0.44	0.44	1.92	0.76
	150	0.05	0.12	0.34	1.07	1.53	0.65
	200	0.07	0.13	0.33	1.06	1.51	0.63
	250	0.06	0.14	0.25	1.07	1.82	0.76
	300	0.06	0.13	0.23	1.28	1.84	0.80

d_{10} = represents 10% of the cumulative droplet frequency; d_{25} = represents 25% of the cumulative droplet frequency; d_{50} = represents the mean cumulative droplet frequency; d_{75} = represents 75% of the cumulative droplet frequency; d_{90} = represents 90% of the cumulative droplet frequency

Table 2.11 Droplet statistical parameter for droplet diameters for different types of nozzles

Nozzle size (mm)	Pressure (kPa)	$\overline{d}$	d_v	d_{50}	SD_D (m)	CV_D
2	100	0.73	3.12	0.45	0.94	119
	150	0.71	2.94	0.36	0.71	87
	200	0.70	2.79	0.35	0.81	107
	250	0.68	2.68	0.37	0.82	124
3	100	0.67	2.71	0.36	0.85	116
	150	0.69	2.09	0.27	0.71	125
	200	0.60	1.93	0.25	8.0	120
	250	0.59	1.68	0.23	0.84	114
4	100	0.78	2.81	0.27	0.84	107
	150	0.76	2.44	0.26	0.68	91
	200	0.73	2.0	0.26	0.71	99
	250	0.72	1.91	0.25	0.79	120
5	100	0.86	2.81	0.34	0.89	106
	150	0.77	2.34	0.24	0.68	91
	200	0.69	2.25	0.23	0.77	114
	250	0.57	2.20	0.21	0.83	147
6	100	0.89	2.80	0.37	1.02	115
	150	0.76	2.79	0.24	0.99	132
	200	0.70	2.19	0.23	0.95	136
	250	0.68	1.49	0.23	0.87	127
7	100	0.80	2.99	0.44	0.92	119
	150	0.79	2.39	0.35	0.67	85
	200	0.75	2.21	0.33	0.71	106
	250	0.66	1.92	0.25	0.79	121

$\overline{d}$ = arithmetic mean droplet; d_v = the volume weighted average droplet diameter; SD_D = the standard deviation; CV_D = is the coefficient of variation

2.4.6 Conclusion

This study evaluated the hydraulic performance of a newly designed dynamic fluidic sprinkler using different types of nozzles at different operating pressures. The following conclusions can be drawn.

The smallest radius of throw was obtained when the sprinkler was operated at the pressure of 100 kPa, while the maximum radius of throw was obtained when the sprinkler was operated at the pressure of 250 kPa. The distance of throw increased with the increase in diameters of nozzle sizes. However, there was no significant different between the radius of throw for 250 and 150 kPa. With the rising cost of energy, it is appropriate to operate under 150 kPa in order to save water.

The comparison of water distribution profiles at different operating pressures showed that all the different nozzle sizes produced parabola-shaped profiles, while the 5.5-mm nozzle size was flatter at a low pressure of 150 kPa. This implies that a 5.5-mm nozzle size can improve the non-uniform water distribution and save water for sprinkler-irrigated fields.

For all the nozzle sizes, 5.5 mm gave the highest computed uniformity value of 86%, at a low pressure of 150 kPa. There was no significant difference between 250 and 150 kPa. Comparatively, the sprinkler with a 5.5-mm nozzle produced a better uniformity, and the average CU obtained was within the acceptable range.

The mean droplet diameter for the nozzles sizes of 2, 3, 4, 5.5, 6 and 7 mm ranged from 0 to 4.2, 0 to 3.7, 0 to 3.6, 0 to 3.2, 0 to 0.5 and 0 to 3.8 mm, respectively. The comparison of the droplet size distribution for the various sizes showed that 5.5 mm had the optimum droplet diameter of 3.2 mm. The largest droplet size had a maximum value of 4.0 for a 2-mm nozzle size. Hence, using a 5.5 mm nozzle size can produce the optimum droplet sizes, which can minimize losses caused by wind drift and evaporation.

References

1. Kahlown MA, Raoof A, Zubair M, Kemper WD (2007) Water use efficiency and economic feasibility of growing rice and wheat with sprinkler irrigation in the Indus Basin of Pakistan. Agric Water Manag 87:292–298
2. Montazar A, Sadeghi M (2008) Effects of applied water and sprinkler uniformity on alfalfa growth and hay yield. Agric Water Manag 95:1279–1287
3. Kincaid DC (1996) Spray drop kinetic energy from irrigation sprinklers. Trans ASAF 39:847–853
4. Kulkarni S (2011) Innovative technologies for Water saving in Irrigated Agriculture. Int J Water Resour Arid Environ 1:226–231
5. Howell TA (2001) Enhancing water use efficiency in irrigated agriculture. Agronomy J 93:281–289
6. Shin S, Bae S (2013) Simulation of water entry of an elastic wedge using the FDS scheme and HCIB method. J Hydrodynamics 25:450–458
7. Darko RO, Yuan SQ, Liu JP, Yan HF, Zhu XY (2017) Overview of advances in improving uniformity and water use efficiency of sprinkler irrigation. Int J Agric Biol Eng 10:1–15
8. Liu JP, Zhu XY, Yuan SQ, Wan JH, Chikangaise P (2018) Hydraulic performance assessment of sprinkler irrigation with rotating spray plate sprinklers in indoor experiments. J Irrig Drain Eng 144. https://doi.org/10.1061/(ASCE)IR.1943-4774.0001333
9. Li J, Rao M (2000) Sprinkler water distributions as affected by winter wheat canopy. Irrig Sci 20:29–35
10. Tang LD, Yuan SQ, Qiu ZP (2018) Development and research status of water turbine for hose reel irrigator. J Drain Irrig Mach Eng 36:963–968
11. Al-Ghobari HM (2006) Effect of maintenance on the performance of sprinkler irrigation systems and irrigation water conservation. Food Sci Agric Res Cent Res Bull 141:1–6
12. Kohl RA, Koh KS, Deboer DW (1987) Chemigation drift and volatilization potential. Appl Eng Agric 3:174–177
13. Dwomoh FA, Yuan SQ, Li H (2014) Sprinkler rotation and water application rate for the newly designed complete fluidic sprinkler and impact sprinkler. Int J Agric Biol Eng 7:38–46

14. Zhu X, Chikangaise P, Shi W, Chen WH, Yuan S (2018) Review of intelligent sprinkler irrigation technologies for remote autonomous system. Int J Agric Biol Eng 11:23–30
15. DeBoer DW, Monnens MJ, Kincaid DC (2001) Measurement of sprinkler drop size. Appl Eng Agric 17:11–15
16. Kohl RA, Bernuth RD, Heubner G (1985) Drop size distribution measurement problems using a laser unit. Trans ASAE 28:190–192
17. Fukui Y, Nakanishi k, Okamura S (1980) Computer elevation of sprinkler uniformity. Irrig Sci 2:23–32
18. Thompson A, Gilley JR, Norman JM (1993) Sprinkler water droplet evaporation a plant canopy model. Trans ASARE 36:743–750
19. Lorenzini G, Wrachien D (2005) Performance assessment of sprinkler irrigation system a new indicator for spray evaporation losses. J Int Commun Irrig Drain 54:295–305
20. Bautista C, Salvador R, Burguete J (2009) Comparing methodologies for the characterization of water drops emitted by an irrigation sprinkler. Trans ASABE 52:1493–1504
21. Wilcox JC, Swailes GE (1947) Uniformity of water distribution by some under tree orchard sprinklers. Sci Agric 27:565–583
22. Hart WE, Reynolds WN (1965) Analytical design of sprinkler systems. Trans Am Soc Agric Eng 8:83–85
23. Kay M (1988) Sprinkler irrigation equipment and practice; Batsford Limited. London, UK
24. Keller J, Bliesner RD (1990) Sprinkle and trickle irrigation; Van Nostrand Reinhold Pun. New York, NY, USA
25. Xiang QJ, Xu ZD, Chen C (2018) Experiments on air and water suction capability of 30PY impact sprinkler. J Drain Irrig Mach Eng 36:82–87
26. Karmeli D (1978) Estimating sprinkler distribution pattern using linear regression. Trans ASAE 21:682–686
27. Liu JP, Liu WZ, Bao Y, Zhang Q, Liu XF (2017) Drop size distribution experiments of gas-liquid two phase's fluidic sprinkler. J Drain Irrig Mach Eng 35:731–736
28. Liu JP, Yuan SQ, Li H, Zhu XY (2013) Numerical simulation and experimental study on a new type variable-rate fluidic sprinkler. J Agric Sci Technol 15:569–581
29. Dwomoh FA, Yuan S, Hong L (2013) Field performance characteristics of fluidic sprinkler. Appl Eng Agric 29:529–536
30. Zhu X, Yuan S, Jiang J, Liu J, Liu X (2015) Comparison of fluidic and impact sprinklers based on hydraulic performance. Irrig Sci 33:367–374
31. Zhang AM, Sun PN, Ming FR, Colagrossi A (2017) Smoothed particle hydrodynamics and its applications in fluid-structure interactions. J Hydrodynamics 29:187–216
32. Zhu X, Yuan S, Liu J (2012) Effect of sprinkler head geometrical parameters on hydraulic performance of fluidic sprinkler. J Irrig Drain Eng 138:1019–1026
33. Zhu X, Fordjour A, Yuan S, Dwomoh F, Ye D (2018) Evaluation of hydraulic performance characteristics of a newly designed dynamic fluidic sprinkler. Water 10(10):1301. https://doi.org/10.3390/w10101301
34. Coanda H (1936) Device for deflecting a stream of elastic fluid projected into an elastic fluid. U.S. Patent No. 2,052,869, 1 Spetember 1936
35. American Society of Biological Enginees (1985) Procedure for sprinkler testing and performance reporting. ASAE S398.1: St. Joseph MI, United States
36. Al-araki GY (2002) Design and evaluation of sprinkler irrigation system. Doctoral Dissertation; University of Khartoum: Khartoum, Sudan
37. Chen D, Wallender WW (1985) Droplet size distribution and water application with low-pressure sprinklers. Trans ASAE 28:511–516
38. Li J, Kawano H, Yu K (1994) Droplet size distributions from different shaped sprinkler nozzles. Trans ASAE 37:1871–1878
39. Tarjuelo JM, Montero J, Valiente M, Honrubia FT, Ortiz J (1999) Irrigation uniformity with medium size sprinkler, part I: characterization of water distribution in no—wind conditions. Trans ASAE 42:677–689

40. King BA, Winward TW, Bjorneberg DL (2013) Comparison of sprinkler droplet size and velocity measurements using a laser precipitation meter and photographic method. Am Soc Agric Biol Eng 131594348. https://doi.org/10.13031/aim.20131594348
41. Sudheera KP, Panda RK (2000) Digital image processing for determining drop sizes from irrigation spray nozzles. Agric Water Manag 45:159–167
42. Liu J, Liu X, Zhu X, Yuan S (2016) Droplet characterization of a complete fluidic sprinkler with different nozzle dimensions. Biosyst Eng 148:90–100
43. Lu J, Cheng J (2018) Numerical simulation analysis of energy conversion in hydraulic turbine of hose reel irrigator JP75. J Drain Irrig Mach Eng 36:448–453

Chapter 3
Numerical Simulation and Experimental Study on Internal Flow Characteristic in the Dynamic Fluidic Sprinkler

Abstract A detailed study of the relationship between velocity distribution, length of tube and nozzle sizes was conducted using a dynamic fluidic sprinkler. Therefore, the objectives of this chapter were to study (1) inner flow characteristics of dynamic fluidic sprinkler. (2) compare numerical simulation and experimental results (3) to introduce an empirical equation of the variation trend of rotation speed for the newly designed sprinkler. A mathematical model for simulation of the inner flow distribution of the sprinkler was obtained by using computational fluid dynamics. The results were validated by numerical simulation and compared to the experimental results. The results revealed that the nozzle diameter, length of the tube and the operating pressure had a significant influence on the rotation speed of the sprinkler. This study provides baseline information to improve water application efficiency for crop production in sprinkler irrigated fields.

Keywords Inner flow · Nozzle · numerical simulation · Water saving

3.1 Introduction

Water-use efficiency (WUE) in crop production is a major problem in the design and management of irrigation systems. Sprinkler irrigation is characterized by high-potential irrigation efficiency [1], and it has been widely used in agriculture for water conservation. The sprinkler head is regarded as a key component of any sprinkler irrigation system, and its hydraulic performance can affect the irrigation efficiency of sprinkler systems. Most studies about sprinkler irrigation have focused on hydraulic performance [2, 3]. An analysis of the flow behavior of the water flow in the sprinkler inner field is also important to understand the micro-characteristics of the sprinkler. Many investigators have applied CFD as a numerical simulation tool to carry out different investigations on sprinkler irrigation. Wang (2006) analyzed the flow characteristics in the emitter used in the drip irrigation inner field using computational fluid dynamics (CFD) techniques. Some basic studies have been performed on the inner flow characteristics of sprinklers. In 1933, the designer at the company Rainbird [4] developed an impact sprinkler for the first time. The inner flow characteristics

X. Zhu et al., *Dynamic Fluidic Sprinkler and Intelligent Sprinkler Irrigation Technologies*, Smart Agriculture 3, https://doi.org/10.1007/978-981-19-8319-1_3

of the sprinkler compartment and pressure losses were analyzed using ANSYS software [5–7]. A 3D turbulent simulation was used to analyze the flow behavior and compared the flow rate, static pressure distribution and kinetic energy values of the sprinkler [8]. Their results indicated that the turbulence model can accurately predict the working pressure and the relationship between the outlet velocities.

In 2005, researchers at Jiangsu University in China developed a new type of fluidic sprinkler. Its principle of operation is based on the "Coanda effect" to perform the function of rotation. Several theoretical, numerical and experimental studies have been conducted to improve the structural and hydraulic performance of the complete fluidic sprinkle. Palfrey and Liburdy [9] studied the main flow characteristic of the turbulent offset jet. Bourque and Newman [10] studied the mean flow characteristic of the wall-attaching offset jet. Hoch and Jiji [11] also researched into a numerical prediction of the jet trajectory and jet reattachment length· Their researcher works revealed that when the primary flow jet becomes reattached to the right side, the pressure in the two sidewalls of the main jet flow exclusively depends on flowing duct length and operating pressure. Song et al. [12] studied the flow and heat transfer characteristics of the 2D jet. Their study demonstrated that the structural optimization approach can be effectively implemented by CFD simulation. Wang and Lu [13] studied the basic theory of wall attachment fluidic; their research was based on only 2D flow. Liu et al. [14] studied numerical and experimental studies on a new type of variable rate of the fluidic sprinkler. Zhu et al. [15] carried out the study on flow characteristics of a wall attaching offset jet in a complete fluidic sprinkler.

They concluded that the results obtained from the numerical simulation could reflect the inner flow of a complete fluidic sprinkler. Dwomoh et al. [16] compared fluidic and impact sprinklers and concluded that variations in quadrant completion times were small for both fluidic and impact sprinklers. Similarly, Hu et al. [17] carried out a study on the fluidic sprinkler and confirmed the need to optimize the structure. Based on their recommendations, a newly designed dynamic fluidic sprinkler was invented by the Research Centre of Fluid Machinery Engineering and Technology, Jiangsu University China. The sprinkler is comparatively cheaper, simple to construct, low consumption of energy and price. No previous studies investigated numerical simulation on dynamic fluidic sprinkler. Therefore, the objectives of this paper were to study (1) inner flow characteristics of the dynamic fluidic sprinkler (2) compare Numerical simulation and experimental results based on rotation speed time using CFD program FLUENT software. (3) introduce the empirical equation of the variation trend of rotation speed for the newly designed sprinkler.

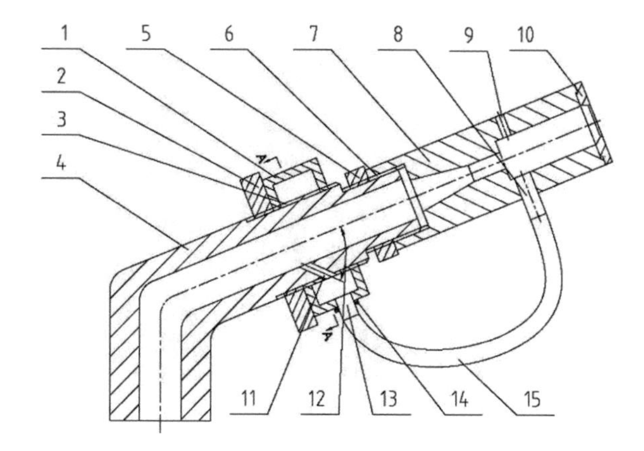

Fig. 3.1 Schematic, pictorial view of the newly fluidic sprinkler head. 1. Water signal tank. 2. First, lock nut. 3. Pipe sprayer 4. Spray body. 5. Second lock nut. 6. Body of the fluidic element. 7. Jet element body. 8. Water inlet. 9. Airhole. 10. Outlet cover plate. 11. Water dividing hole. 12. α degree. 13. Signal nozzle. 14. Third lock nut. 15. Conduit

3.2 Materials and Methods

3.2.1 Design of Newly Dynamic Fluidic Sprinkler Type

Figure 3.1 presents the structure of a dynamic fluidic sprinkler. The profile of the fluidic element was defined by the inner contraction angles, the offset length, and the working area. A prototype of the dynamic fluidic sprinkler was self-designed and locally machined by using a wire-cut electric discharge machining process. The manufacturing tolerance for the size was within 0.02 mm as shown in Fig. 3.1. The working theory of dynamic fluidic sprinkler is based on the theory of the Coanda effect. Water flows from the diameter into the action zone, and the main jet is ejected from the central circular hole. The left and the right sides of the main jet are separated from each other, and the air at both ends is out of circulation. By opening the reverse air hole of left allows air to be pumped into the left cavity. The air gap between the exit at the right side of the element and the water jet is filled by air, such that the pressures on both sides of the main jet are basically equal. At the same time, the nozzle receives the signal water on the left edge of the water jet, then the signal water flows in the tube to the inlet signal [18–20].

3.2.2 Design of the Nozzles

Most sprinklers have two nozzles, the main nozzle and an auxiliary nozzle that discharge water in the form of a jet into the air. Nozzles with different diameters are the principal parameters that affect the hydraulic performances of sprinklers. Circular nozzles are commonly used on fluidic sprinklers due to the advantages of the large spray range and simple configurations and convenience to machine [21].

(a) dynamic fluidic sprinkler (b) nozzles

Fig. 3.2 A prototype of the dynamic fluidic sprinkler and nozzle sizes

Therefore, the test nozzles were self-designed and locally machined using a wire-cut electric discharge machining (EDM) process. The prototypes of the nozzles are shown in Fig. 3.2. The inlet diameter of the nozzle was set as 15 mm, while the outlet diameters were chosen as 3, 4, 5, 6, and 7 mm.

3.2.3 Numerical Simulation

Dynamic fluidic sprinkler which consists of a fluidic element and contraction angle was employed in this study. The three- dimensional model of the dynamic fluidic sprinkler is illustrated in Fig. 3.3. The commercial code ANSYS fluent 19.2 was adopted to conduct the numerical calculation. The full flow fields were divided into different fluid domains and each of them was meshed by unstructured hexahedron mesh via FLUENT CFD. This sprinkler works under variable pressure because of the pressure-adjusting device, that is, the inlet of the sprinkler does not keep a fixed pressure. As can be seen in Fig. 3.4, in the variable-rate fluidic sprinkler, one of the moving inserts is stationary and is referred to as the static insert. The pressure at the different rotating angles of the sprinkler was adjusted. The movement insert and static insert were in the rotating angle of 0 to 90 degrees and the black part was the flow cross-section. The movements insert rotated deasil and the angle increased as the sprinkler rotates. For square spraying shape, the varied flow cross-section from 0º to 90º is the same with 90º to 180º, from 180º to 270º is the same with 90º to 180º, from 0º to 180º is the same with 270º to 360º. In ANSYS fluent under the reference frame, the moving insert was set as frame and mesh motion, the rotational velocity $\omega = 0.07$ rad/s was obtained from a plot of the moment against angular velocity and under the relative to a cell, the zone was set as a rotor. Equation (3.6) was written in user-defined function (UDF) to calculate the rotation speed.

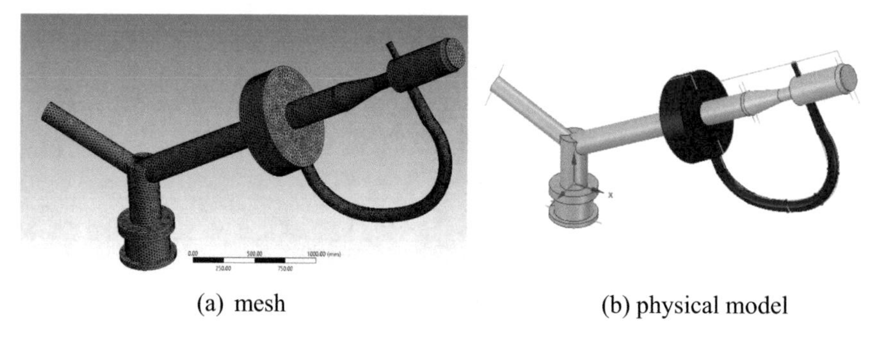

(a) mesh (b) physical model

Fig. 3.3 3D model and unstructured mesh

Fig. 3.4 Schematic diagram of static and movement insert of the sprinkler

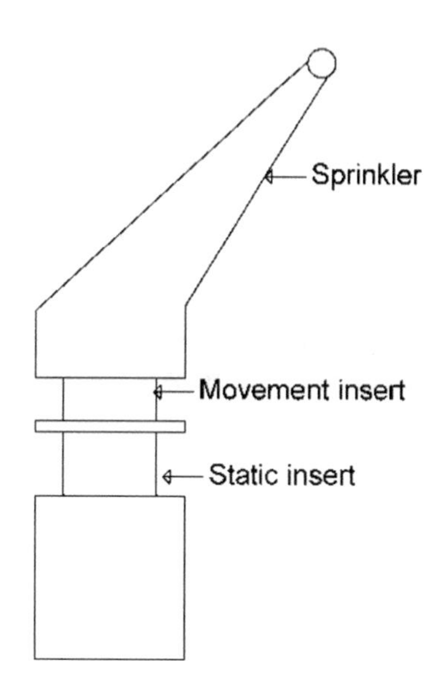

3.2.3.1 Mathematical Model

The momentum equation, shown below, is dependent on the volume fractions of all phases through the properties ρ and μ.

$$\frac{\partial}{\partial t}(\rho\vec{v}) + \nabla.(\rho\vec{v}\vec{v}) = -\nabla\rho + \nabla.\left[\mu\left(\nabla\vec{v} + \nabla\vec{v}^{T}\right)\right] + \rho\vec{g} + \vec{F} \tag{3.1}$$

The energy equation as shown in the equation below.

$$\frac{\partial}{\partial t}(\rho E) + \nabla.\left(\vec{v}(\rho E + \rho)\right) = \nabla.\left(k_{eff}\nabla T\right) + S_h \tag{3.2}$$

where E = VOF model treats energy, T = temperature as mass-averaged variables and can be expressed as Eq. (3.3)

$$\mathrm{E} = \frac{\sum_{q=1}^{n} a_q p_q}{\sum_{\rho=1}^{n} a_q p_q E_q} \tag{3.3}$$

where E_q for each phase is based on the specific heat of that phase and the shared temperature.

The properties ρ and k_{eff} (effective thermal conductivity) are shared by the phases. The source term S_h contains contributions from radiation, as well as any other volumetric heat sources.

The FLUENT models selected are standard k -ε, RNG k -ε, realizable k -ε and so on. The working principle of the fluidic sprinkler is based on gap-liquid flow. The Governing equations for unsteady, three dimensional and viscous flow is given as:

$$\partial^{(\rho u_i \varepsilon)}/\partial \chi_j = \frac{\partial}{\partial \chi_j} + \left(a_\kappa \mu_{eff} \frac{\partial}{\partial \chi_j}\right) + \mu_i s^2 - \rho\varepsilon$$

$$\mu_{eff} = \mu + \mu_i s = \sqrt[2]{2 S_{ij} S_{ij}} \tag{3.4}$$

where μ_{eff} = effective viscosity, α_κ = Prandtl number in k equation pulsation rate equation (ε equation) in the model, R = source item created by deformation rate

$$\partial \frac{(\rho u_i^\varepsilon)}{\partial \chi_i} = \frac{\partial}{\partial \chi_i}\left(a_\varepsilon u_{eff} \frac{\partial \varepsilon}{\partial \chi_i}\right) + c_{i\varepsilon} \mu_t S^2 \frac{\varepsilon}{\kappa} - C_{2\varepsilon} \rho \frac{\varepsilon^2}{\kappa} - R \tag{3.5}$$

3.2.3.2 Rotation Speed

The rotation speed of the sprinkler can be written as specified equation below.

$$\mathrm{KE}_R = \frac{1}{2} \mathrm{I} w^2 Q,\ w = \frac{d\theta}{dt},$$

$$\mathrm{KE}_R = \frac{1}{2} \mathrm{I} \left(\frac{d\theta}{dt}\right)^2 Q$$

$$\frac{2\mathrm{KE}_R}{\mathrm{I}Q} = \left(\frac{d\theta}{dt}\right)^2$$

$$\sqrt{\frac{2\mathrm{KE}_R}{\mathrm{I}Q}} = \frac{d\theta}{dt}$$

$$dt = \frac{d\theta}{\sqrt{\frac{2\mathrm{KE}_R}{\mathrm{I}Q}}} \tag{3.6}$$

where I = moment of initial; KE_R = rotational kinetic energy; $d\theta$ = angular displacement; dt = time; M = mass of the sprinkler

Coefficient (Q) is given as

$$Q = \frac{3}{4}\left(\frac{Cd}{r}\right)\left(\frac{\rho a}{\rho w}\right) \tag{3.7}$$

where Cd = the frictional resistance and is assumed to be 0.44; ρa = Air density (1.29 kg m^3); r = radius of the sprinkler r nozzle (7.5 mm), ρw = Water density (1 × 103 kg m^3). Therefore when dt is known, the rotation speed can be calculated.

3.2.3.3 Grid Sensitivity Analysis

The grid generation is the next step required to perform after geometry creation. It is one of the tedious and time-consuming parts. To solve the Navier- Stokes equation numerically, the commercial code ANYS 19.2 workbench was adopted to conduct the numerical calculation. The full flow fields were divided into different fluid domains and each of them was meshed by unstructured hexahedron mesh via the CFD. An independence test was performed to ensure the independence of a solution with mesh size as shown in Fig. 3.5. The variation in velocity was found to be negligible after 1.35 million lakh elements compared to the other grid elements, hence the grid size of 1.35 million lakh was chosen to conduct the simulation.

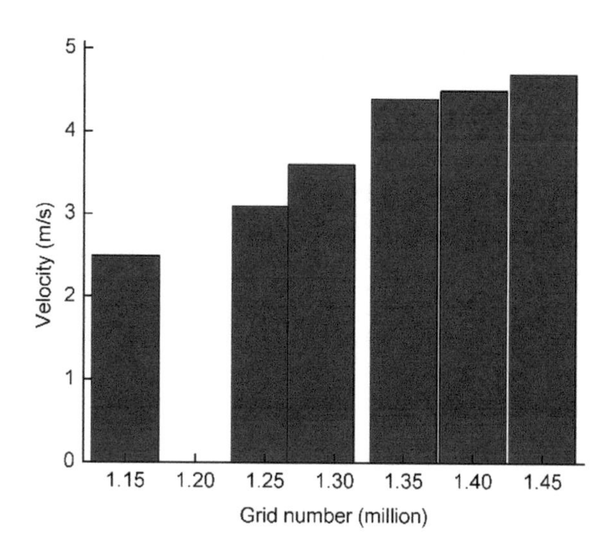

Fig. 3.5. Grid independence check

Table 3.1 Boundary conditions of the sprinkler

Rotation angle (°)	Pressure (kPa)	Nozzle size (mm)	Length of a tube (mm)
90	100	3	15
180	150	4	20
270	200	5	25
360	250	6	30
	300	7	

3.2.3.4 Boundary Conditions

Generally, boundary conditions are needed to be specified on all surfaces of the computation domain. The inlet condition was as pressure, and the outlet condition was set by the outflow. All physical surfaces of the sprinkler were set as non -slip walls. Standard walls functions were used at the solid wall, air and water were used as the working fluid. The convergences precession of the calculation was set at 0.0001. Table 3.1 presents the parameters used in the numerical simulation and experiment. The reason for using these parameters was to compare the performance of each parameter and select the best combination for the sprinkler.

3.2.3.5 Experimental Procedure

The materials used for the experiment include; centrifugal pump, electromagnetic flow meter, and piezometer, valve and dynamic fluidic sprinkler. The sprinkler was installed at a height of 2 m from the ground level with nozzle an elevation angle of 23°. The riser was at an angle of 90° to the horizontal from which was 0.9 m above the ground. Water was pumped from the reservoir through the main pipe and sprayed out from the nozzle as shown in Fig. 3.6. The working pressure was measured by a pressure gauge at the base of the sprinkler had an accuracy of 0.4%. To standardize the laboratory conditions, the sprinkler was run for 20 min before performing the actual experiments. The experiment aimed to determine quadrant completion times for each nozzle and droplet size. Both nozzle and working pressures were all within the manufacturer's recommendations and were varied from100 ~ 300 kPa. To determine the rotation uniformity from one quadrant to another, the total area of rotation was divided into four quadrants and each experiment lasted for 1 h. Quadrants QI, Q2, Q3 and Q4 comprised of radial lines 0, 90; 90, 180; 180, 270 and 270, 0, respectively.

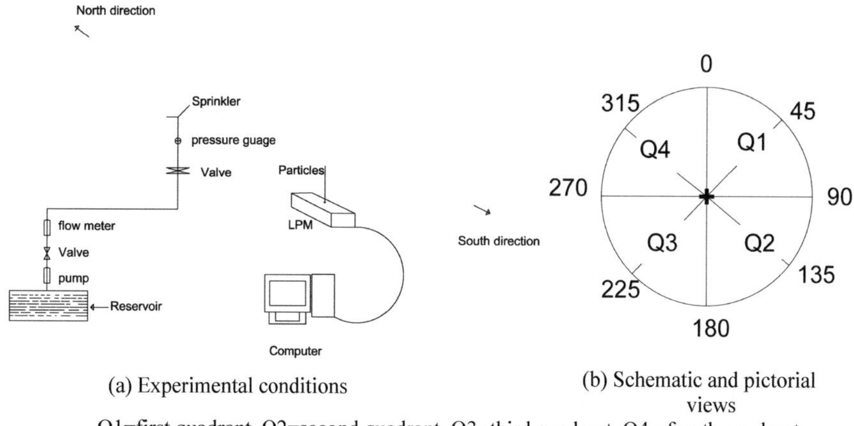

(a) Experimental conditions

(b) Schematic and pictorial views

Q1=first quadrant, Q2=second quadrant, Q3=third quadrant, Q4= fourth quadrant

Fig. 3.6 The layout of cans around the sprinkler in the indoor laboratory

To investigate the sprinkler rotation speed uniformity, a stopwatch was used to record the time taken by the sprinkler to move from one quadrant to the other, in the course of the one-hour duration. Three replications of completion time through the quadrants were recorded for each sprinkler-nozzle-pressure configuration. Drop sizes were measured using a Thies Clima Laser Precipitation Monitor (LPM) was from Adolf Thies GmbH and Co. KG, Gottingen, Germany. Measurements were conducted indoors with no wind at a pressure of 100, 150, 200, 250 and 300 kPa. The LPM measures drop sizes from 0.125 mm to 6.0 mm. Drop size measurements were grouped into 0.1 mm increments (±0.05 mm) for analysis starting with 0.25 mm continuing to 5.95 mm. Measured drops less than 0.2 mm in diameter were discarded as they represent less than 0.01% of the total volume of drops measured. The sprinkler nozzle was located 2 m above the laser beam of the LPM. Measurements were collected at the middle (6 m) and end (12 m) positions of radial distances from the sprinkler. Droplet size parameters used in this study include Dv = average volume diameter and (d, mm) = arithmetic mean diameter as shown in Eqs. (3.8) and (3.9)

$$\overline{d} = \frac{\sum_{i=1}^{n} m_i d_i}{\sum_{i=1}^{n} m_i} \tag{3.8}$$

$$d_v = \frac{\sum_n^n - = 1\, {d_i}^4}{\sum_n^n = 1\, {d_i}^3} \tag{3.9}$$

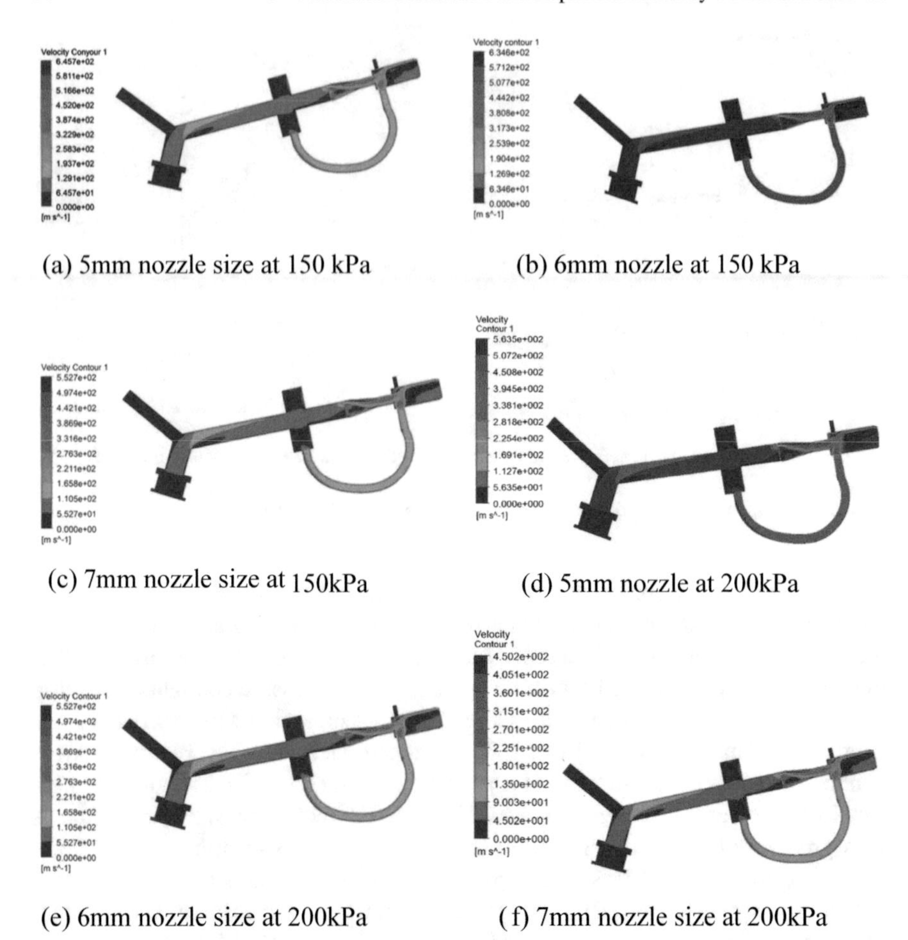

(a) 5mm nozzle size at 150 kPa

(b) 6mm nozzle at 150 kPa

(c) 7mm nozzle size at 150kPa

(d) 5mm nozzle at 200kPa

(e) 6mm nozzle size at 200kPa

(f) 7mm nozzle size at 200kPa

Fig. 3.7 Relationship between velocity and nozzle sizes

3.3 Results and Discussion

3.3.1 Relationship Between Velocity Distribution and Nozzle Sizes

Numerical simulations were carried out to identify velocity distribution within the dynamic fluidic sprinkler with different types of nozzle sizes under a pressure of 150 and 200 kPa. For both pressures velocity distribution decreased steadily as the nozzle sizes increased. Generally, low-velocity gradients existed in 7 mm nozzle size, which improved when the pressure was increased.

There was some flow distortion observed very close to the fluidic element for all the nozzle sizes. This however improved as the nozzle size was increased. The maximum velocity occurred around the fluidic element. This phenomenon is caused by a reduction in pressure on the inner side of the jet due to the entrainment of fluid by the jet. This supports already established results from earlier research works. The 5 mm nozzle size, however, had relatively higher velocity distribution along the surface with a 7 mm nozzle having lower distribution. At the pressure of 200 kPa, the trend was the same. The 5 mm nozzle size saw much improvement in velocity distribution with a low-pressure region. This means that a 5 mm nozzle size could be ideal for field application to maximize the range under the given pressure while avoiding the cost of crop production in sprinkler irrigated fields (Fig. 3.7).

3.3.2 Relationship Between Velocity and Length of the Tube

The relationship between the velocity distribution, length of the tube, and nozzle diameter of 5 mm under different working pressure is shown in Fig. 3.8.

There was some distortion observed close to the inlet of tube length especially 15 mm. Distortion reduced as the length of the tube was increased. Velocity distribution within the tube was much better especially with the 25 mm length of the tube. It was found that the distortion on the 25 mm length of the tube is smaller compared with the other length of the tube. This means that by using a 25 mm tube length, the test sprinkler can achieve better rotation speed. The 15 mm length had a relatively lower velocity distribution along the surface. As pressure was increased velocity also increased, meanwhile velocity distribution was not improved compared with 150 kPa. A possible reason could be that more pressure is been built within the tube resulting in low velocity. It can, therefore, be concluded that the test sprinkler with 5 mm nozzle size, 25 mm tube length, and 150 kPa combination will work efficiently from a water distribution perspective.

3.3.3 Comparison of the Numerical Simulation, Calculated and Experimental Results

An empirical equation for the variation trend of the rotation speed of different nozzles, length of the tube and operating pressures were generated by using curve fitting. The proposed equation is sufficiently accurate and simple to use for rotation speed. According to the equation, the nozzle diameter, length of the tube and the working pressure had a significant influence on the rotation speed of the test sprinkler. Comparing the fitted values, simulated and experimental data, the relative error (er) was less than 4% which means the analysis of the relationship between rotation speed and nozzle diameters, length of tube and pressure is accurate.

$$Sp = 11.5 D_2^{0.001} L^{-0.99} P^{0.1} \tag{3.10}$$

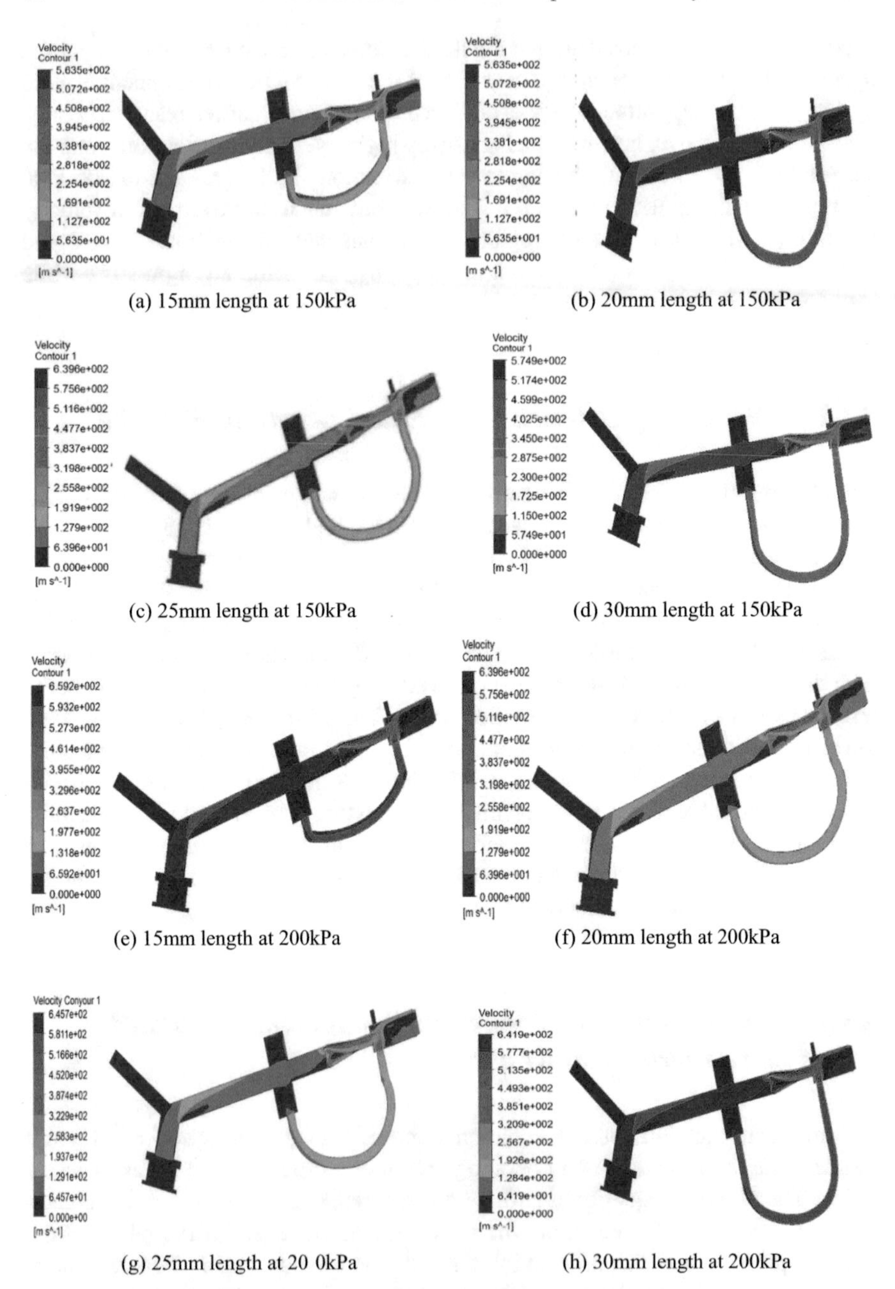

(a) 15mm length at 150kPa (b) 20mm length at 150kPa

(c) 25mm length at 150kPa (d) 30mm length at 150kPa

(e) 15mm length at 200kPa (f) 20mm length at 200kPa

(g) 25mm length at 20 0kPa (h) 30mm length at 200kPa

Fig. 3.8 Relationship between velocity and length of the tube

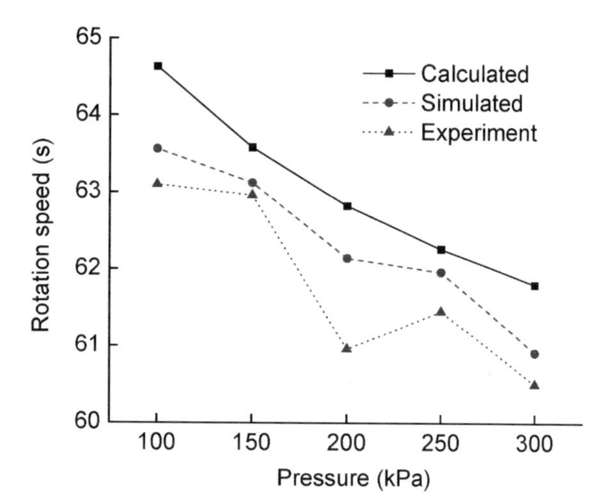

Fig. 3.9 Comparison of the numerical simulation and experimental results

where *Sp* is the rotation speed (seconds), *D*2 is the nozzle diameter (mm), *L* length of the tube (mm) and *p* is the working pressure (kPa).

Computational results from numerical simulations and experiments were matched with the calculated shown in Fig. 3.9 and Table 3.2. From the simulated results, 5 mm nozzle size and 25 mm tube length were found to be ideal. Hence, 5 mm nozzle size and 25 mm tube length were selected for further testing.

From the study, it can be seen that simulated, calculated and experimental dropped gradually as the pressure was increased, and the trend between the calculated and simulated was quite similar. The experimental results were; however, lower than simulated and calculated results. The relative errors at 150 kPa were 0.73, 0.90 and 0.25 simulated, calculated and experiment, respectively. The computational results were relatively higher than experimental results and comparatively lower than the calculated results. This could be attributed to the fluctuation of pressure and external factors. The minimum average deviation between the calculated, simulation and experiment was 0.25%, whiles the maximum was 0.90%. Notwithstanding, the deviations were minimal and therefore the numerical results agreed well with the experimental results and this affirmed the reliability of the empirical equation been generated.

3.3.4 Comparison of Rotation Speed and the Nozzle Sizes

Rotation time measures the period of sprinkler rotation. This can be performed by recording the rotation speed of the sprinkler by using a stopwatch. A fast rotation rate causes a stream jet to bend and creates gaps in the pattern due to friction. However, the slow rate increases application intensity so long as the jet stream remains situational. The repetition of spray intensity floods soil surface and instantaneously prevents air

Table 3.2 Comparison of the numerical simulation, calculated and experimental results for 5.5 mm nozzle size

Pressure (kPa)	Simulated (s)	Calculated (s)	Relative error	Calculated (s)	Experiment (s)	Relative error	Simulated (s)	Experiment (s)	Relative error
100	63.56	64.63	1.68	64.63	63.1	2.42	63.56	63.1	0.73
150	63.12	63.58	0.73	63.58	62.96	0.98	63.12	62.96	0.25
200	62.14	62.82	1.09	62.82	60.96	3.05	62.14	60.96	1.94
250	61.96	62.26	0.49	62.26	61.45	1.32	61.96	61.45	1.48
300	60.91	61.8	0.02	61.8	60.5	2.14	60.91	60.5	0.68

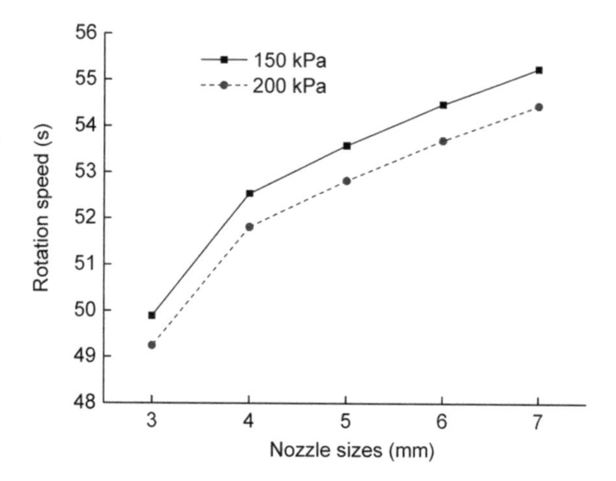

Fig. 3.10 Comparison of rotation speed and nozzle sizes

escape from soil media resulting in surface run-off and sheet erosion. Figure 3.10 shows the relationship between the rotation speed and the nozzle sizes at 150 and 200 kPa.

From the study, it was revealed that the rotation speed increases approximately linearly as the nozzle size increases. As the pressure increases the rotation speed decreases for all the nozzles size. It was found that rotation speed was, however, smaller with smaller nozzle diameters and this is because smaller diameters have greater velocities which make the sprinkler rotate faster. It also appears that when the nozzle was smaller than 5 mm, the rotation speed reduced below 55 s, which means that sprinklers might not operate efficiently. When the nozzle size was too large, the sprinkler tends to give higher rotation speed. This is because less pressure is been built within the sprinkler resulting in low rotation speed. Variations in rotation speed were 14.5, 13.8, 14.1 s, and 14.9 s for 3, 4, 5, 6, and 7 mm, respectively. A comparison of nozzle sizes demonstrated that 5 mm nozzle size gave optimum rotation stability compared with other types of nozzles used in this study. These results are slightly better than those obtained by.

3.3.5 Relationship Between Rotation Speed and Length of the Tube

Figure 3.11 shows the relationship between the rotation speed and the length of the tube at 150 and 200 kPa. In this study, a 5 mm nozzle was maintained whiles the pressure and length of the tube were varied. From the graph, it can be seen that as the length of the tube increases rotation speed also decreases. This could be attributed to a loss of pressure in the length of the tube resulting in low rotation speed.

Comparatively, 200 kPa decreased slightly as the length of the tube increased. With an increase in tube length, 25 mm appears to produce optimum results compared with

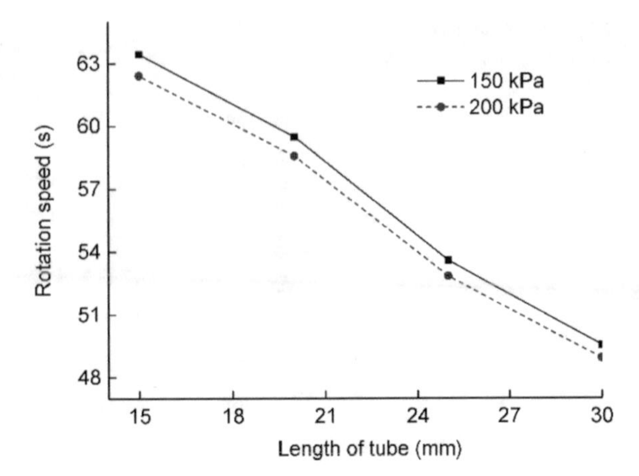

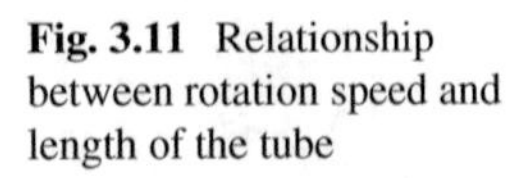
Fig. 3.11 Relationship between rotation speed and length of the tube

the other length of tubes. It was found that the differences in rotation speed were insignificant ($p > 0.05$). From the graph, it can be observed that the length of the tube had a slight influence on the rotation speed of the test sprinkler.

3.3.6 *Effect of Internal Velocity Distribution on Hydraulic Performance*

The shape of the water distribution pattern, uniformity, range, droplet size, and velocities is mainly determined by the sprinkler model and its internal design, discharge angle and jet breakup mechanism. The relationship between internal velocity distribution, a distance of throw and droplet size with different nozzles diameter under a different working pressure of 150 and 200 kPa is shown in Fig. 3.12. The distance of throw increased with an increase in pressure. This is possible because, at a high-velocity condition, the jet breaks up quickly, resulting in a higher radius of throw.

It can be observed from the figure that the velocity distribution increased as the size of the droplets decreased for both pressures. For both pressures, optimum velocity was obtained at 5 m-s and thereafter decreased as the velocity increases, which can maximum losses caused by wind drift and evaporation. These results are similar to those obtained by previous researchers who used different sprinkler types. Using a 5 mm nozzle size will produce optimum droplet sizes, which can fight wind drift and evaporation losses. This is because large droplets possess high kinetic energy, and on impact, they disrupt the soil surface, especially soils with crustiness problems, leading to sealing of the soil surface.

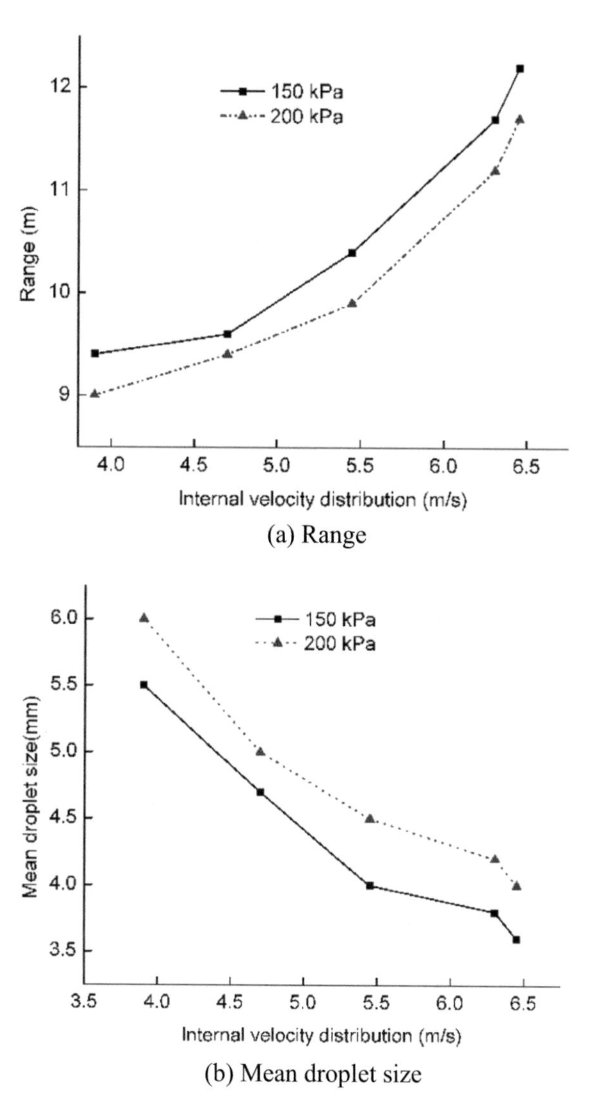

Fig. 3.12 Relationship between velocity, range and droplet size

References

1. Clemmens AJ, Dedrick AR (1994) Irrigation techniques and evaluations. Adv Ser Agric Sci 14:64–103
2. Hendawi M, Molle B, Folton C, Granier J (2005) Measurement accuracy analysis of sprinkler irrigation rainfall in relation to collector shape. J Irrig Drain Eng 131:477–483
3. Sourell H, Faci JM, Playan E (2003) Performance of rotating spray plate sprinklers indoor experiments. J Irrig Drain Eng 129:376–380
4. Li J, Kawano H (1995) Simulating water-drop movement from noncircular sprinkler nozzles. J Irrig Drain Eng 121(6):152–158

5. Yan HJ, Jin HZ (2004) Study on the discharge coefficient of non-rotatable sprays for center-pivot system. J Irrig Drain 23:55–58
6. Yan HJ, Liu ZQ, Wang FX, Yang XG, Wang M (2007) Research and development of impact sprinkler in China. J China Agric Univ 12:77–80
7. Yuan SQ, Zhu XY, Li H, Ren ZY (2005) Numerical simulation of inner flow for complete fluidic sprinkler using computational fluid dynamics. Trans CSAM 36(10):46–49
8. Yan HJ, Ou YJ, Kazuhiro Xu CB (2009) Numerical and experimental investigations on internal flow characteristic in the impact sprinkler. Irrig Drain Syst 23:11–23
9. Palfrey JR, Liburdy JA (1986) Mean flow characteristic of turbulent offset jet. Trans ASME 108:82–88
10. Bourque L, Newman BG (1960) Reattachment of two-dimension incompressible jet to an Adjacent flatplate. Aeronaut Q 11:201232. https://doi.org/10.1017/S0001925900001797
11. Hoch J, Jiji LM (1981) Two dimensions turbulent offset jet-boundary interaction. Trans ASME 103:154–161
12. Song HB, Yoon SH, Lee DH (2000) Flow and heat transfer characteristic of a two-dimensional oblique wall attaching offset jet. Int J Heat Mass Transf 43:2395–2404
13. Wang Q, Lu Y (1996) Theoretical laws for determining the main parameters of a wall-attachment jet flow. J Jilin Agric Univ 18:69–71
14. Liu JP, Yuan SQ, Li H, Zhu XY (2013) Numerical simulation and experimental study on a new type variable-rate fluidic sprinkler. J Agric Sci Technol 15(3):569–581
15. Zhu X, Yuan S, Liu J (2012) Effect of sprinkler head geometrical parameters on the hydraulic performance of fluidic sprinkler. J Irrig Drain Eng 138(11):1943–4774
16. Dwomoh FA, Shouqi Y, Hong L (2014) Sprinkler rotation and water application rate for the newly designed complete fluidic sprinkler and impact sprinkler. Int J Agric Biol Eng 7(4):38–46. http://doi10.9790/1684-11467073
17. Hu G, Zhu XY, Yuan SQ, Zhang LG, Li YF (2019) Comparison of ranges of fluidic sprinkler predicted with BP and RBF neural network models. J Drain Irrig Mach Eng. 37(3):263–269
18. Li H, Yuan SQ, Liu JP, Xiang QJ, Zhu XY, Xie FQ (2007) Wall-attachment fluidic sprinkler. Ch. Patent No. 101224444 B
19. Liu JP, Liu XF, Zhu XY, Yuan SQ (2016) Droplet characterization of a complete fluidic sprinkler with different nozzle dimensions. Biosyst Eng 65(4):2–529
20. Zhu XY, Fordjour A, Yuan SQ, Dwomoh F, Ye DX (2018) Evaluation of hydraulic performance characteristics of a newly designed dynamic fluidic sprinkler. Water 10, 1301. https://doi.org/10.3390/w10101301
21. Han S, Evans RG, Kroeger MW (1994) Sprinkler distribution patterns in windy conditions. Trans ASAE 37(5):1481–1489

Chapter 4
Effect of Riser Height on Rotation Uniformity and Application Rate of the Dynamic Fluidic Sprinkler

Abstract The objective of this chapter was to investigate the effect of riser heights on rotation uniformity and application rate of the newly designed dynamic fluidic sprinkler. The Dynamic fluidic sprinkler was tested using different nozzle sizes of 5 and 6 mm. It was found that the effect of riser height had a significant effect on application rate and uniformity.

Keywords Rotation · Nozzle · Uniformity coefficient · Riser height

4.1 Introduction

Performance characterization has been relevant to project and manage irrigation systems. System performance can be defined as the degree to which the system's products and operation meet the need of their users and the efficiency with which the system uses resources at its disposal [1–4]. The sprinkler irrigation performance could be tested by various performance indicators. The number of indicators is influenced by the level of detail in which the device operator wants to quantify performance [5–7] (Liu et al. 2016). Over the years, researches have acknowledged sprinkler rotation speed variation as the major factor influencing the overall uniformity of water distribution. Excessive rotational speed will lead to jet bending. This leads to the gap in the pattern, which is often repeated in the same area due to the friction characteristics [8–10]. A speed that is too slow would cause an increase in the intensity of application for the time the jet stream remains in one area. This intensity when repeated will cause sealing of the soil which may lead to run off and sheet erosion [11, 12]. Extensive research works have been focused on improving the structure and hydraulics performance of complete fluidic sprinklers [12–15, 16, 17] (Liu et al. 2016; Xu et al. 2019). From their studies, it was found that it is necessary to redesign the fluidic structure of the fluidic sprinkler to enhance the rotation stability and minimize the inconsistency in the water application rates. Based on their findings, a newly designed dynamic fluidic sprinkler was developed by the Research Centre of Fluid Machinery Engineering and Technology, Jiangsu University China. It has the following advantages, easy to construct, low loss of energy, low price and this type of sprinkler is suitable for fruits and vegetables. Therefore, the objective of this paper

X. Zhu et al., *Dynamic Fluidic Sprinkler and Intelligent Sprinkler Irrigation Technologies*, Smart Agriculture 3, https://doi.org/10.1007/978-981-19-8319-1_4

was to study the influence of riser height on the rotational uniformity and application rate of the newly sprinkler under different working pressures and nozzle types.

4.2 Materials and Methods

The Dynamic fluidic sprinkler was tested using different nozzle sizes of 5 and 6 mm. The basis of using these nozzles size is to select the best combination. The dynamic fluidic sprinkler was inverted by Jiangsu University, P.R China. The working principle of dynamic fluidic is based on the Coanda effect theory. Water is sprayed from the main nozzle to the work area, a region of the low-pressure eddy is formed on both sides of the working area. The left and the right sides of the main jet are separated from each other, and the air at both ends is out of circulation. By opening the reverse air hole of the left allows air to be pumped into the left hole. The air gap between the exit at the right side of the element and the water jet is filled by air, such that the pressures on both sides of the main jet are basically equal. The main flow jet is bent toward the boundary and eventually attached to it because the left pressure is much larger than the right pressure. This phenomenon is repeated step by step, and the sprinkler rotates gradually to the other side [18, 19, 20] (Liu et al. 2018).

4.2.1 Experimental Procedures

The study on the effect of riser height on rotation uniformity and application rate was performed in the laboratory of the Research Centre of Fluid Machinery Engineering and Technology, Jiangsu University, P.R. China. The laboratory is circular—shaped with 44 m in diameter and 18 m in height. A Centrifugal pump was used to supply water from reservoirs with constant water levels. The experiment was carried out in indoor facilities to ensure uninterrupted radial water distribution and avoid water loss and drift [21–23, 24] (Liu et al. 2013). The sprinkler head was mounted on a 1.3, 1.5, 1.7 and 1.9 m riser with a horizontal direction of 90 degrees and was placed about 0.9 m above the ground. The following working pressures 150 and 200 kPa were tested. In order to standardize the environmental conditions, the sprinkler was operated for several minutes. Catch cans used in performing the experiments were cylindrical in shape, 200 mm in diameter and 600 mm in height. In order to perform an orthogonal pattern test, the catch cans were arranged in eight radial lines around the sprinkler as shown in Fig. 4.1a. Figure 4.1b shows the experimental conditions in the indoor laboratory. Figure 4.1c shows the photo of the dynamic fluidic sprinkler. Each line consisted of 10 catch cans at a distance of 1 m around the sprinkler. The total area of rotation was divided into four quadrants in order to determine the rotation uniformity from one quadrant to others. Each quadrant consisted of 3 radial lines. Quadrants Q1, Q2, Q3 and Q4 comprised of radial lines 0, 45, 90; 90, 135, 180; 180, 225, 270 and 270, 315, 0, respectively. The sprinkler flow rates

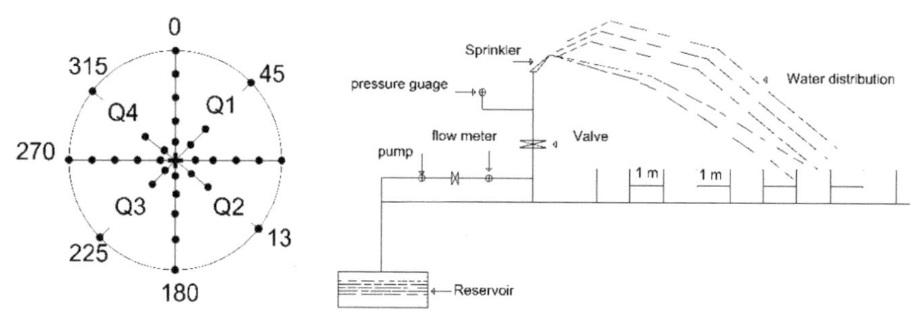

(a) Schematic and pictorial views (b) Experimental conditions

(c) Photo of the dynamic fluidic sprinkler

Fig. 4.1 The layout of cans around the sprinkler in the indoor laboratory

were controlled by pressure regulation. The working pressure at the bottom of the sprinkler was regulated and maintained by the valve through a pressure gauge with accuracy less than (+1%). All the nozzles size and operating pressures were within the manufacture's recommendations. Measurement of application depth was carried out in accordance with the ASABE S398.1 (American Society of Agricultural and Biological Engineers (ASABE), 1985). The experiment lasted for an hour, and the water depth in the catch cans was measured with a measuring cylinder. A stopwatch was used to determine the time taken by one sprinkler to move from one quadrant to another. The processes were repeated three times for each nozzle and pressure. The time for a complete revolution of the sprinkler head was similarly recorded. Water application depths in the catch cans and sprinkler rotation speed through each quadrant and for a complete rotation of the sprinkler head were recorded.

4.2.2 Evaluation of Sprinkler Performance

Sprinkler performance evaluation regarding coefficients of uniformity, water application, the standard deviation of application rate and quadrant time was determined from the experimental data for each of the sprinkler nozzles size.

Standard deviations were performed from the experiments for each of the sprinkler nozzles configuration by using Eq. 4.1.

$$SD = \sqrt{1 \Big/ (n-1) \sum_{1=1}^{n} \left(t - \bar{t}\right)^2} \tag{4.1}$$

where, *SD* is the standard deviation; t_i is the mean quadrant completion time through the *i*th quadrant in a complete rotation (360°); *n* is the number of replications and is the mean of completion time through the four quadrants.

Christiansen's Coefficient of uniformity was obtained by Eq. 4.2.

$$CU = 100\left(1 - \frac{\sum_{i=1}^{n} |X_i - \mu|}{\sum_{i=1}^{n} X_i}\right) \tag{4.2}$$

where n = number of catch cans; x_i = application depth, mm; μ = average depth, mm; and CU = coefficient of uniformity, %

4.2.3 Overlap Water Distribution

A program was written to simulate the quadrants overlapped in both horizontal and vertical directions as shown in Eqs. 4.3 and 4.4, respectively.

For horizontal overlap of quadrants, was coded in matrixes as follows:

$$F = [C_{il}\ O_{il}\ O_{il}] + [O_{il}\ b_{il}\ O_{il}] \tag{4.3}$$

For the vertical overlap of quadrants, the resultant pattern is as follows:

$$G = \begin{bmatrix} C_{il} \\ O_{ij} \\ O_{ik} \end{bmatrix} + \begin{bmatrix} O_{il} \\ b_{ij} \\ O_{ik} \end{bmatrix} + \ldots\ldots. \tag{4.4}$$

where, *bij* and *Cil* are matrices, with the catch can readings in the quadrants as elements; *Oik* and *Oij* are null matrices inserted for purposes of mathematical correctness. The null matrices technically indicate areas where no water was applied.

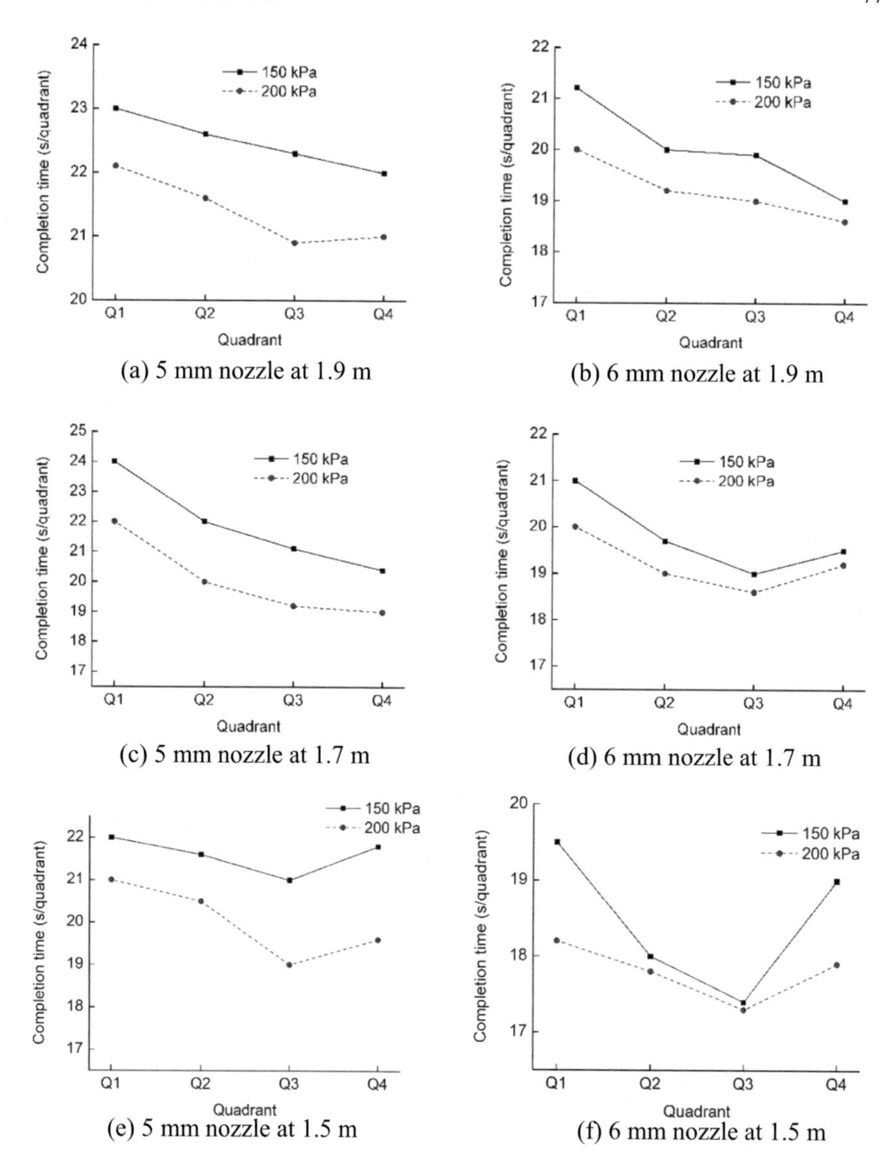

Fig. 4.2 Quadrant completion time for different risers at different operating pressures and nozzle sizes

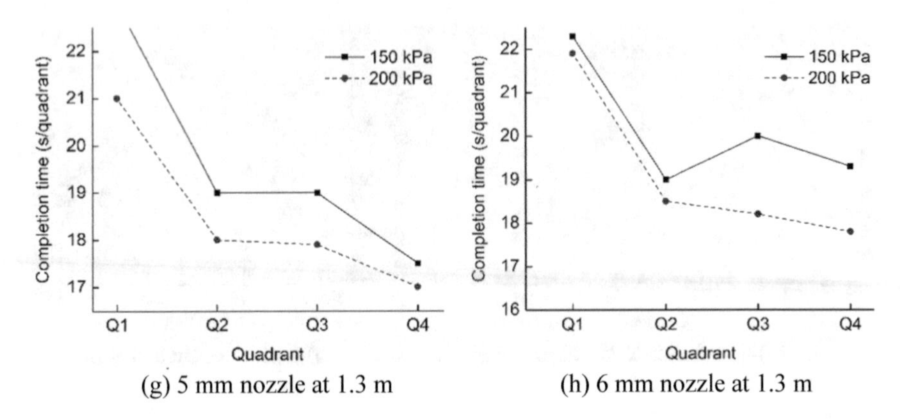

(g) 5 mm nozzle at 1.3 m (h) 6 mm nozzle at 1.3 m

Fig. 4.2 (continued)

4.3 Results and Discussion

4.3.1 Quadrant Completion Time

Figure 4.2 presents the results of quadrant completion time for different risers under different operating pressures and nozzle diameters. Generally, for the same pressure, increasing the outlet diameter of the nozzles resulted in a decreased in the speed of rotation. For the same nozzle diameter, increasing the pressure resulted in a decreased speed rotation. This is in agreement with the findings of [25, 26]. The variations in quadrant completion time at different risers showed that 1.9 m risers were lower compared with the other riser.

The changes in quadrant completion times were 23, 22.7, 22.6 and 22 s, respectively, Q1, Q2, Q3, and Q4, as shown in Fig. 4.2a. It was found that the differences in quadrant completion times were insignificant and decreases approximately linearly with nozzle size 5 mm, which means that using 1.9 m riser gives optimum rotation stability compared with the other types of risers used in this study. The comparison of nozzle sizes under different working pressures shows that the nozzle size of 5 mm had better completion time. The range of quadrant completion times was 22–23, 19–21.2, 21–23 and 17.5–23 s for 1.9, 1.7, 1.5 and 1.3 m, respectively. The obtained results for the quadrant completion time were better than previous findings by Dwomoh et al. (2014) [25]. In the case of the pressure, the variation completion time of 150 kPa was, however, smaller compared with 200 kPa for all the risers, measured ranges of quadrant completion time of 20 ~ 23 and 17 ~ 22 s were recorded at operating pressures of 150 and 200 kPa, respectively.

4.3.2 Deviation in Water Application Intensity

Figure 4.3 presents the differences in standard deviations of water application depth across the radial lines within the four quadrants of rotation for different risers under different operating pressures and nozzle sizes. The standard deviations resulting from 200 kPa at different riser heights were much larger. The deviations ranged from 0.16 ~ 0.28, 0.19 ~ 0.5, 0.2 ~ 0.59, 0.22 ~ 0.74 mm h^{-1} for 1.9, 1.7, 1.5 and 1.3 m, respectively. The largest with a value of 0.74 mm h^{-1}, this occurrence can be attributed to a reduction of riser height. The differences in standard deviations were much lower for all the nozzle sizes under 150 kPa. The deviation ranged from 0.14 ~ 0.26, 0.14 ~ 0.42, 0.18 ~ 0.45, 0.2 ~ 0.66 mm h^{-1} considering the height of 1.9, 1.7, 1.5 and 1.3 m, respectively. With respect to the nozzle sizes, 5 mm gave better standard deviations and the deviations ranged from 0.14 ~ 0.26, 0.15 ~ 0.45, 0.2 ~ 0.7, 0.23 ~ 0.74 mm h^{-1} at radial heights of 1.9, 1.7, 1.5 and 1.3 m, respectively. This is caused by the pressure variations created by the alignment of signal nozzles and fluid flow through the signal pipes in the fluidic component. In the case of the 6 mm nozzle size, large deviations were recorded. The deviations ranged from 0.2 ~ 0.5, 0.14 ~ 0.55, 0.2 ~ 0.7, 0.23 ~ 0.74 mm h^{-1} at radial heights of 1.9, 1.7, 1.5 and 1.3 m, respectively. It can be concluded that the test sprinkler with a 5 mm nozzle size was working efficiently from the perspective of water distribution uniformity.

It was found in Fig. 4.3 that the effect of riser height was very strong, the lowest deviation of water application intensity was 0.14 ~ 0.26 in Fig. 4.3a, the highest deviation of water application intensity was 0.16 ~ 0.74 in Fig. 4.3p. The comparison at different risers showed that differences in standard deviations were much lower for 1.9 m riser height. The deviation ranged from 0.14 ~ 0.26, 0.14 ~ 0.42, 0.18 ~ 0.50, 0.13 ~ 0.69 mm h^{-1} considering the riser height of 1.9, 1.7, 1.5 and 1.3 m, as shown in Fig. 4.33a, e, i and n. Although the deviations were quite small between 1.9 and 1.7 m, with an average coefficient of variation (CV) of 9.2%, and 1.7 m was 13.2%. A significant difference was found for 1.3 m, while no significant difference occurred for 1.9 m between the pressure of 150 and 200 kPa.

4.3.3 Comparison of Water Distribution Profiles

Figure 4.4 shows plots of average water application patterns in the quadrants at different risers under different operating pressures and nozzle sizes at 150 and 200 kPa, respectively. As can be seen in Fig. 4.4, the application rates were generally higher near the sprinkler for the nozzles with 6 mm diameter, while that from the nozzle with 5 mm diameter was much better. As presented in Fig. 4.4a, using sprinkler with 5 mm as an example, the application rates increased to 7.1 mm h^{-1} at 7 m away from the sprinkler and then decreased to 3.4 mm h^{-1} at 13.3 m from the sprinkler under a pressure of 150 kPa. With respect to 6 mm nozzle size, the application rates increased to 6.6 mm h^{-1} at 5 m away from the sprinkler and then

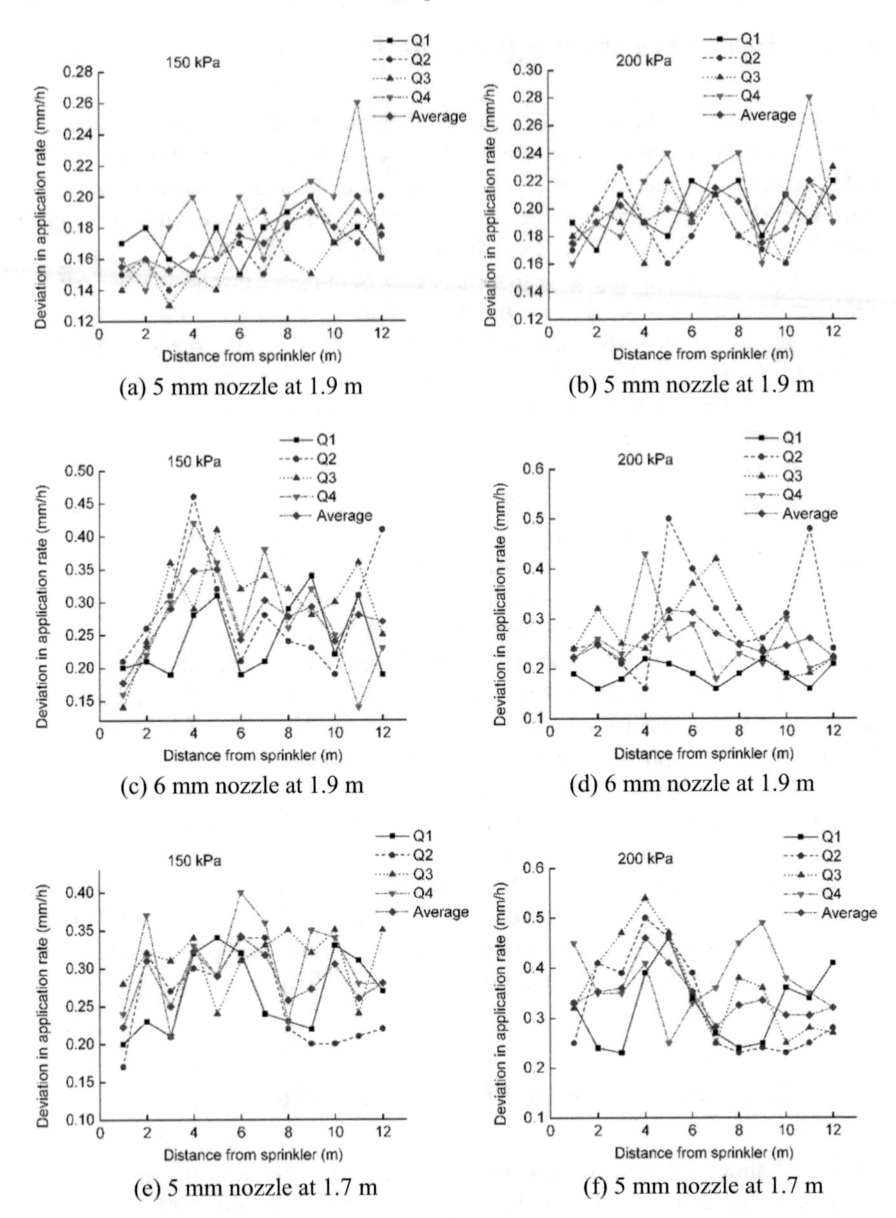

(a) 5 mm nozzle at 1.9 m (b) 5 mm nozzle at 1.9 m

(c) 6 mm nozzle at 1.9 m (d) 6 mm nozzle at 1.9 m

(e) 5 mm nozzle at 1.7 m (f) 5 mm nozzle at 1.7 m

Fig. 4.3 Standard deviations of water application depth across the radial lines within the four quadrants at different risers under different operating pressures and nozzle diameters

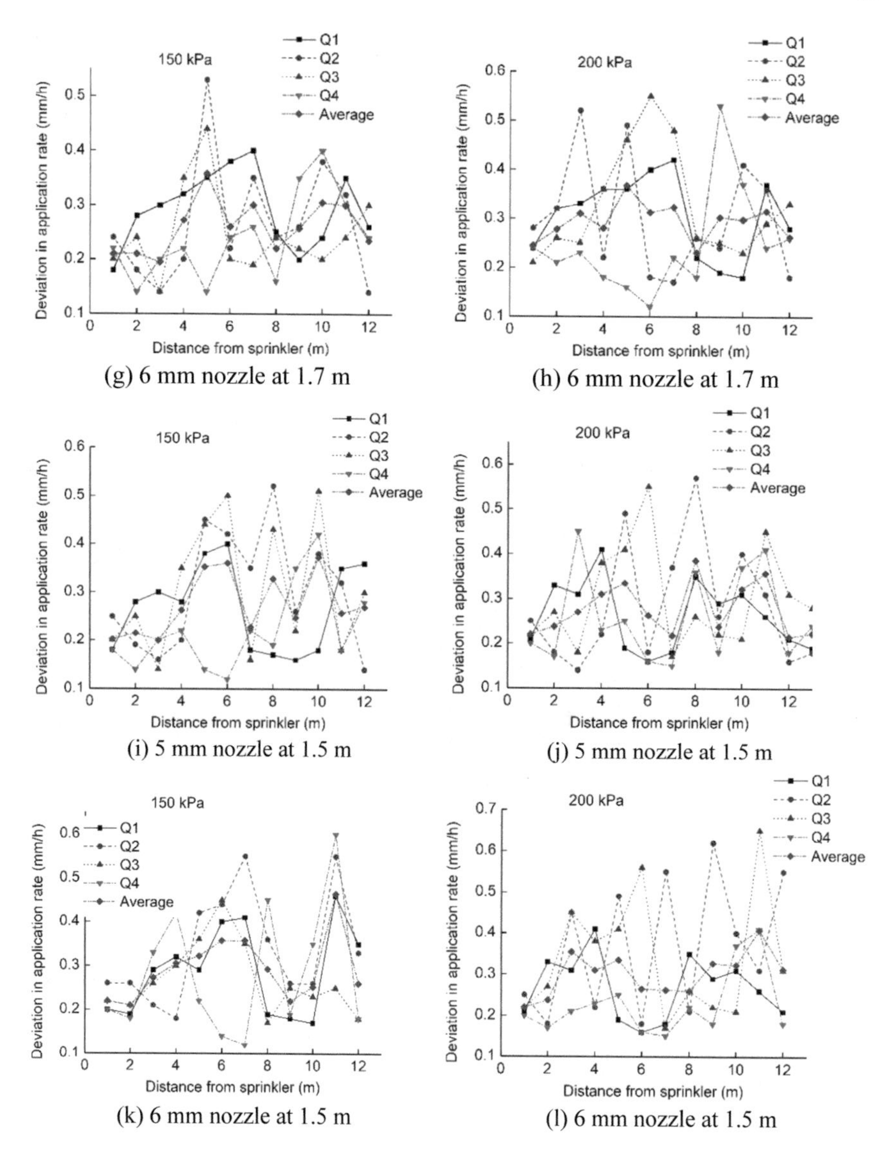

(g) 6 mm nozzle at 1.7 m

(h) 6 mm nozzle at 1.7 m

(i) 5 mm nozzle at 1.5 m

(j) 5 mm nozzle at 1.5 m

(k) 6 mm nozzle at 1.5 m

(l) 6 mm nozzle at 1.5 m

Fig. 4.3 (continued)

decreased to 2.71 mm h^{-1} at 13.3 m from the sprinkler under a pressure of 150 kPa. This means that sprinklers with a 5 mm nozzle could be ideal for field applications to maximize the range under the given pressure. It was found that the nozzles with 5 mm diameter contributed less water near the sprinkler, and the application profiles were doughnut-shaped when the nozzle outlet diameter was increased. More water was applied near the sprinkler when the 5 mm nozzle was used which was due to greater interruption of the nozzle.

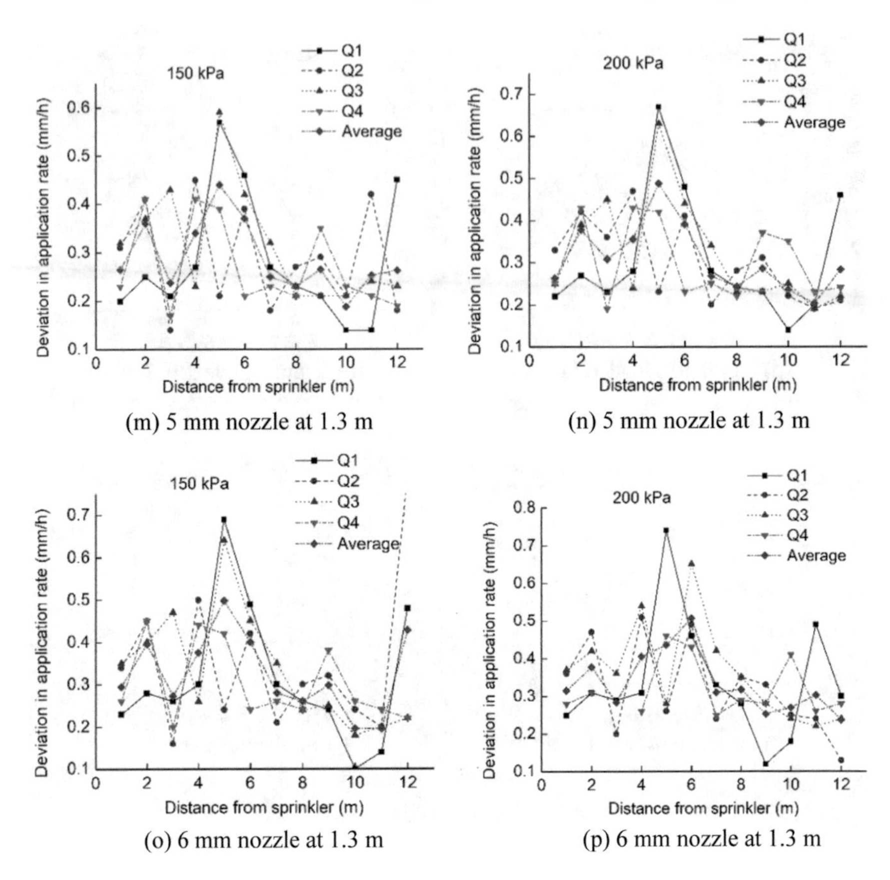

(m) 5 mm nozzle at 1.3 m

(n) 5 mm nozzle at 1.3 m

(o) 6 mm nozzle at 1.3 m

(p) 6 mm nozzle at 1.3 m

Fig. 4.3 (continued)

Application rates from different riser heights showed that water distribution was improved when the riser height was increased. Application rates from 1.7 m increased to a maximum of 6.8 mm h^{-1} at 6 m from the sprinkler and decreased to a minimum of 2.84 mm h^{-1} under a pressure of 150 kPa. Riser height of 1.9 m had a maximum application rate of 6.7 mm h^{-1} at 4 m from the sprinkler, which decreased gradually to 3.3 mm h^{-1}. Meanwhile, the application rates from 1.7 m riser were improved near the sprinkler with the highest value of 5.98 mm h^{-1} at 5 m from the sprinkler, which gradually decreased to 2.74 mm h^{-1}. A possible reason could be that when the inner contraction angle was large, the internal turbulent flow was less uniform from the nozzle outlet and more water was applied near the sprinkler. Relating rotation time in the quadrants to average water application intensity pattern in their corresponding quadrants, the quadrants which recorded the highest rotation times also registered the highest average water application intensities for most portions along the radial lines and vice versa. From Fig. 4.4, the order of increasing rotation time is (Q1, Q2, Q3, and Q4). Meanwhile, the effect of working pressure on the application rate was

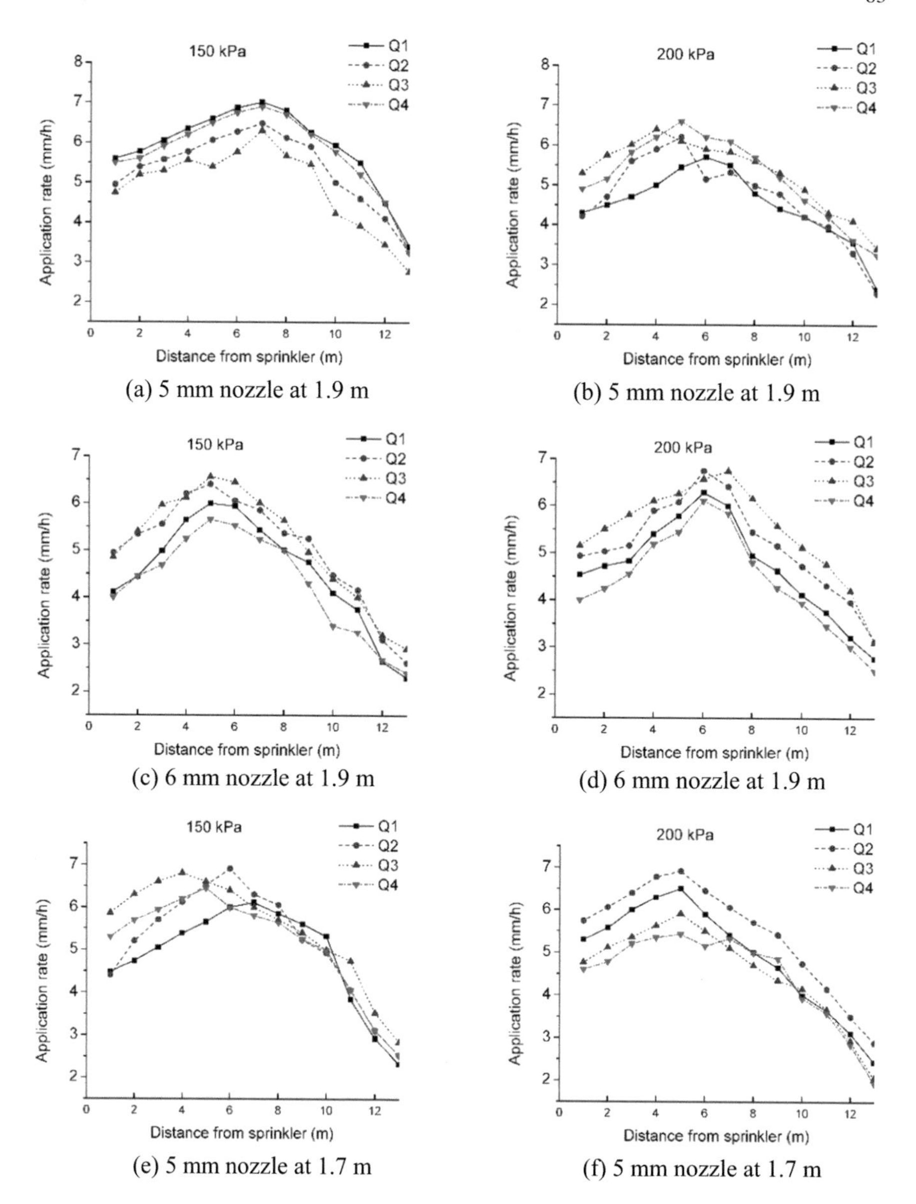

(a) 5 mm nozzle at 1.9 m

(b) 5 mm nozzle at 1.9 m

(c) 6 mm nozzle at 1.9 m

(d) 6 mm nozzle at 1.9 m

(e) 5 mm nozzle at 1.7 m

(f) 5 mm nozzle at 1.7 m

Fig. 4.4 Plots of average water application patterns in the quadrants at different risers under different operating pressures and nozzle sizes at 150 and 200 kPa

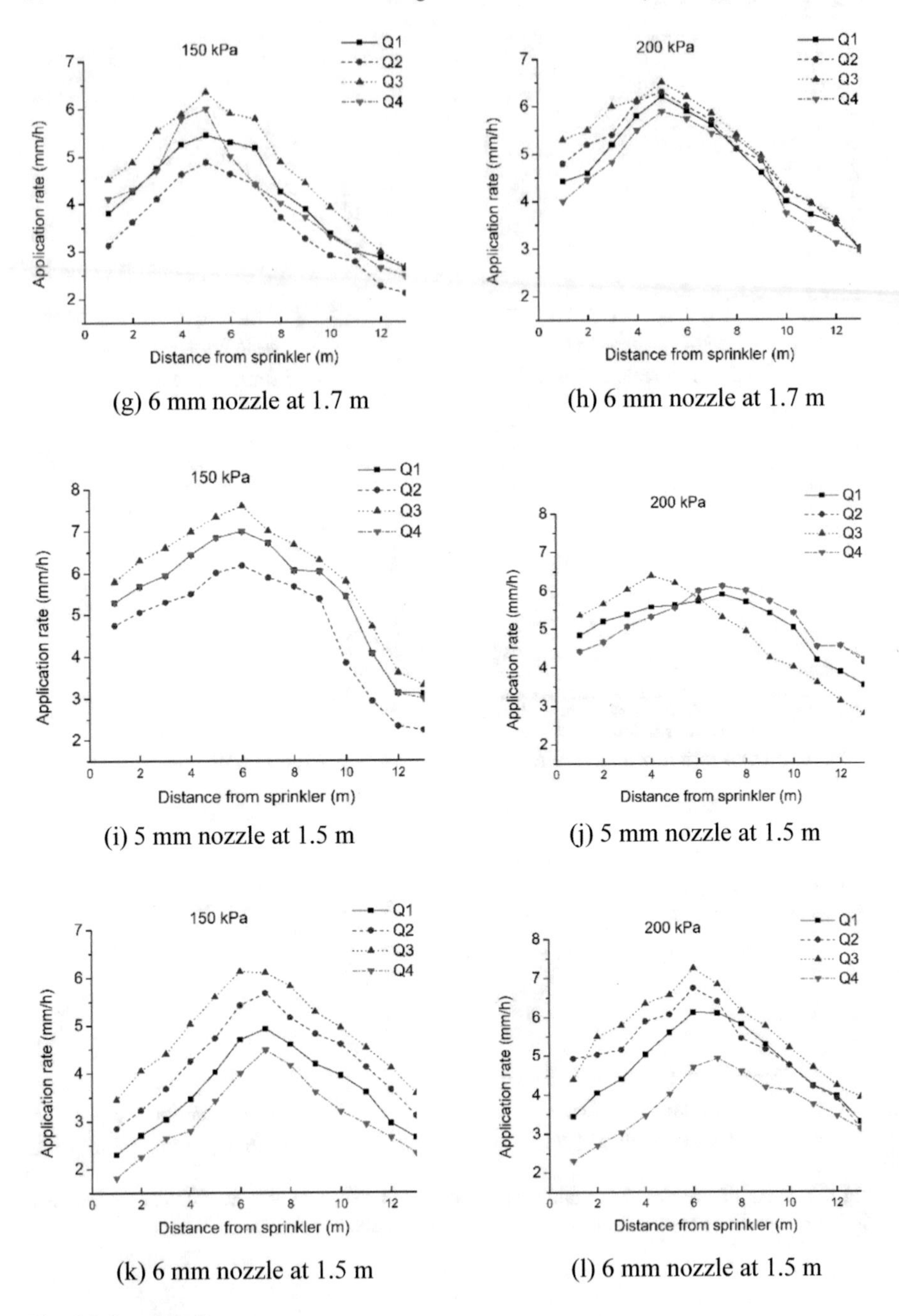

Fig. 4.4 (continued)

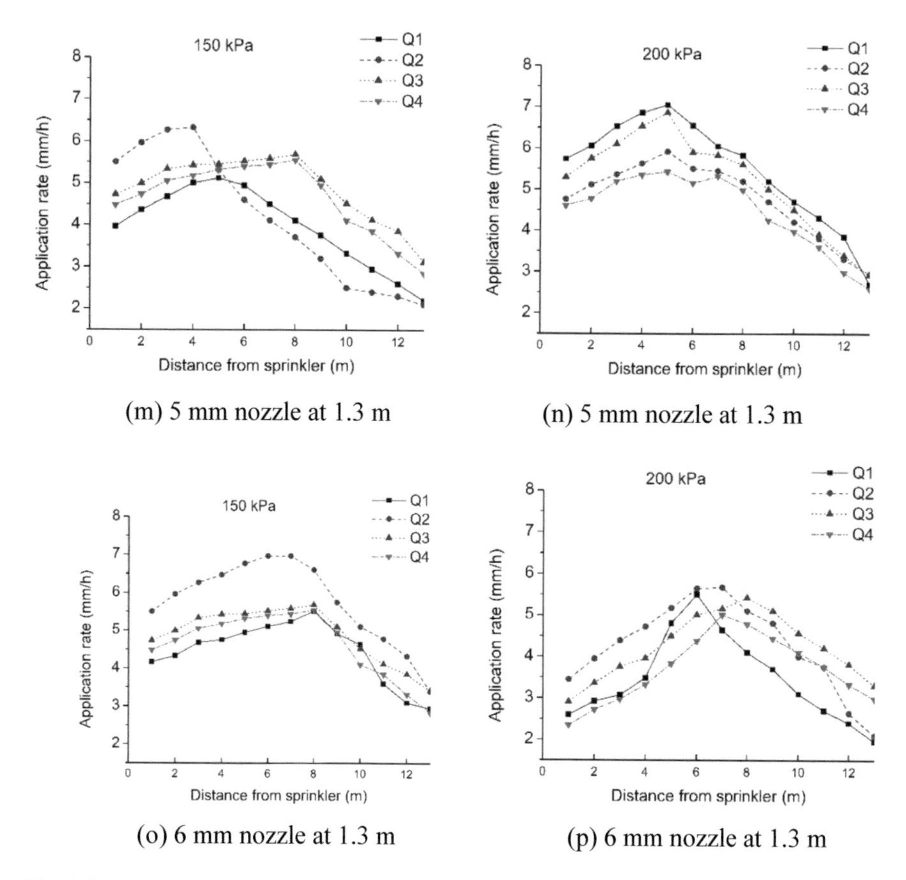

Fig. 4.4 (continued)

minimal compared to that of the nozzle diameter and riser height. This conclusion is similar to previous results obtained for the combined effect of nozzle diameter and pressure on the water distribution for irrigation sprinklers. The comparison shows that 5 mm can be used together with the 1.9 m riser height to improve the water distribution of the fluidic sprinkler under low-pressure conditions.

From the above analysis, among the riser heights, 1.9 m with the 5 mm nozzle size at the pressure of 150 kPa performed better. A further test was carried out with 150 kPa to find out the water distribution for each quadrant. Figure 4.5 presents the plots of quadrant water distribution for 150 kPa at an installation height of 1.9 m above ground level. It can be observed that variations in the contour and colour maps around the sprinkler had different application rates. Rings that have similar colour indicate uniform water distribution pattern, whiles different colours in the ring represent non-uniform water distribution patterns. From the plots, it can be seen that uniformity of water distribution was quite uniform at 1.9 m height. This means that the nozzle size of 5 mm can improve the uneven distribution of water and save

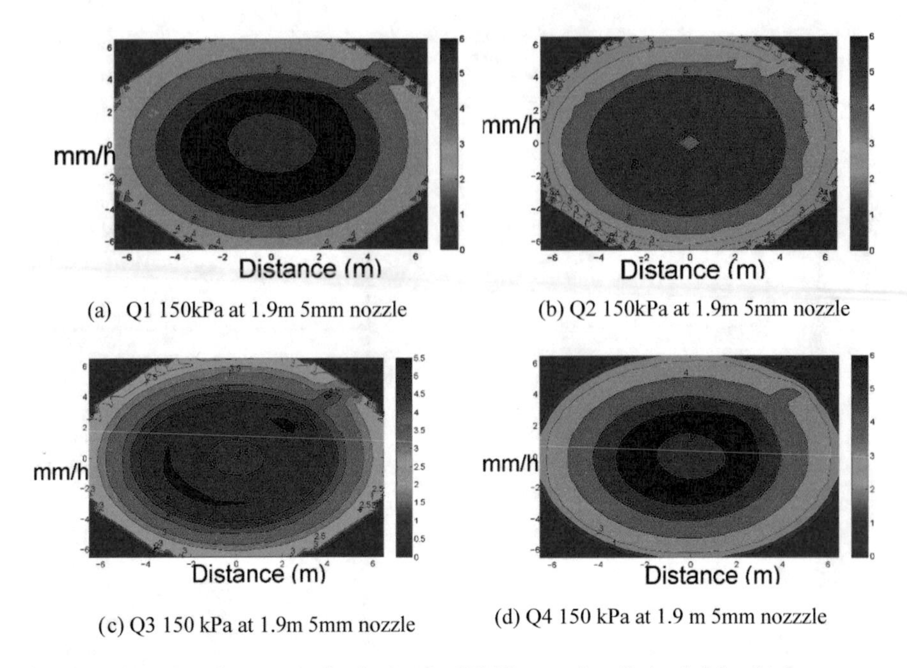

(a) Q1 150kPa at 1.9m 5mm nozzle

(b) Q2 150kPa at 1.9m 5mm nozzle

(c) Q3 150 kPa at 1.9m 5mm nozzle

(d) Q4 150 kPa at 1.9 m 5mm nozzzle

Fig. 4.5 Plots of quadrant water distribution for 150 kPa at an installation height of 1.9 m

water for crop production. As a consequence, differences in water distribution can be seen in most areas around the sprinkler in the case of the other height tested in this study, which is in agreement with the results of variations in rotations speed with respect to the quadrants (Fig. 4.6).

4.3.4 Overlap Distribution Analysis

Figure 4.7 illustrates the overlapped CUs for different risers under different operating pressures and nozzle sizes. For all the cases, the CUs values obtained from the overlapping quadrant were higher than those obtained from CU for individual quadrants. As the distance from the sprinkler increased, the coefficient of uniformity also increased until it got to the maximum and then decreased for all the pressures and nozzles. These results are in agreement with already established results from earlier research works [14, 27, 28].

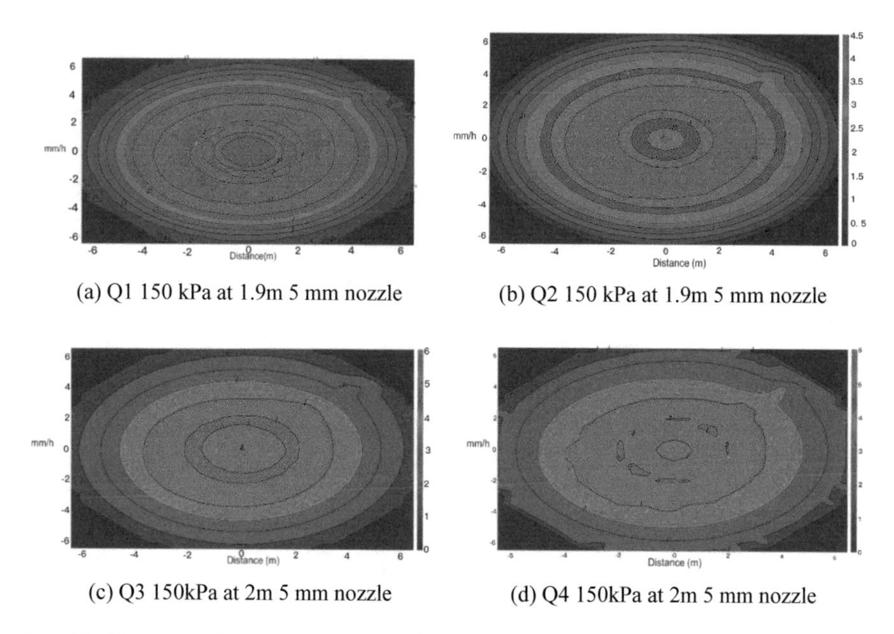

(a) Q1 150 kPa at 1.9m 5 mm nozzle (b) Q2 150 kPa at 1.9m 5 mm nozzle

(c) Q3 150kPa at 2m 5 mm nozzle (d) Q4 150kPa at 2m 5 mm nozzle

Fig. 4.6 Plots of quadrant water distribution for 150 kPa at an installation height of 1.9 m

It can be found in Fig. 4.7a, overlapped CUs from 5 mm nozzle at a height of 1.9 m under pressure of 150 kPa was much higher compared with other ones. The range of CUs values for overlapped quadrants was as follows: 78% at a spacing of 10 to 73% at 80% spacing of overlapped. The highest occurred at 50% spacing and increased with overlapped spacing of 10 to 50%, ranging from 78 to 87% with an average of 82.5%; subsequently, the overlapped uniformity decreased with spacing from 50 to 80%, the CU value ranged from 84 to 73% with an average of 78.5% at a height of 1.9 m. The difference in CU values for overlapped quadrants was 2% for Q3 and at 20% spacing and the lowest was 2% for Q2 at 20% spacing. As shown in Fig. 4.7a, the positions with higher water distributions overlapped the positions with low water distributions from the next quadrant, giving more uniform water distributions, and the other way around was also true considering the same sprinkler spacing. This phenomenon is caused by higher water application depths overlapped with positions of lower application depth from the other quadrant, producing overall uniform distribution. The sprinkler with the 1.9 m riser height gave small differences in CUs of 2.9% at Q1 and Q3 at 40% spacing, and the lowest was 0.18% at Q1 and Q3 at 20% spacing in the overlapped uniformity. This means that by using the 1.9 m riser height, the test sprinkler can be placed in fewer rows to achieve a better overlap

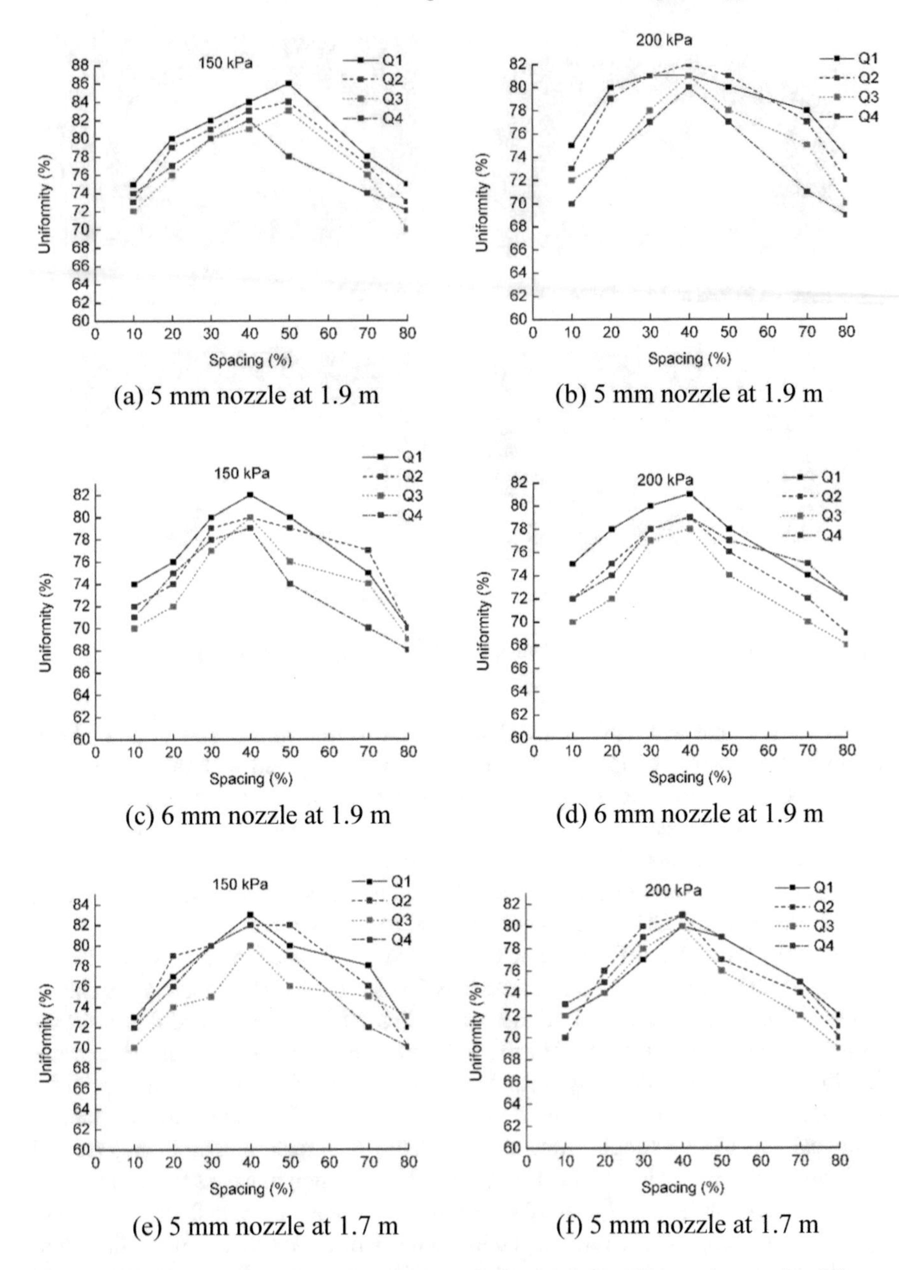

Fig. 4.7 CUs values of overlapped quadrants against spacing for different risers under different operating pressures and nozzle sizes

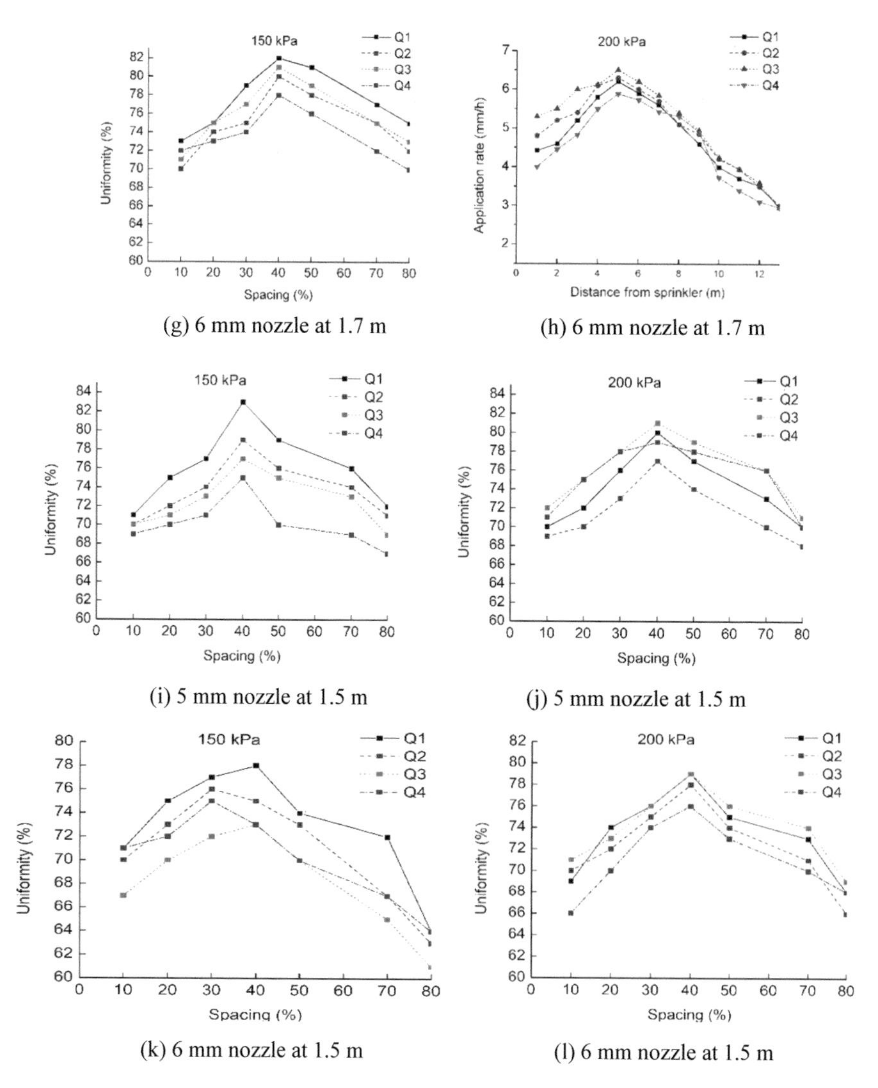

(g) 6 mm nozzle at 1.7 m

(h) 6 mm nozzle at 1.7 m

(i) 5 mm nozzle at 1.5 m

(j) 5 mm nozzle at 1.5 m

(k) 6 mm nozzle at 1.5 m

(l) 6 mm nozzle at 1.5 m

Fig. 4.7 (continued)

water distribution and uniformity on sprinkler irrigated fields. By using the working pressure of 150 kPa gave a better overlap of water distribution and uniformity.

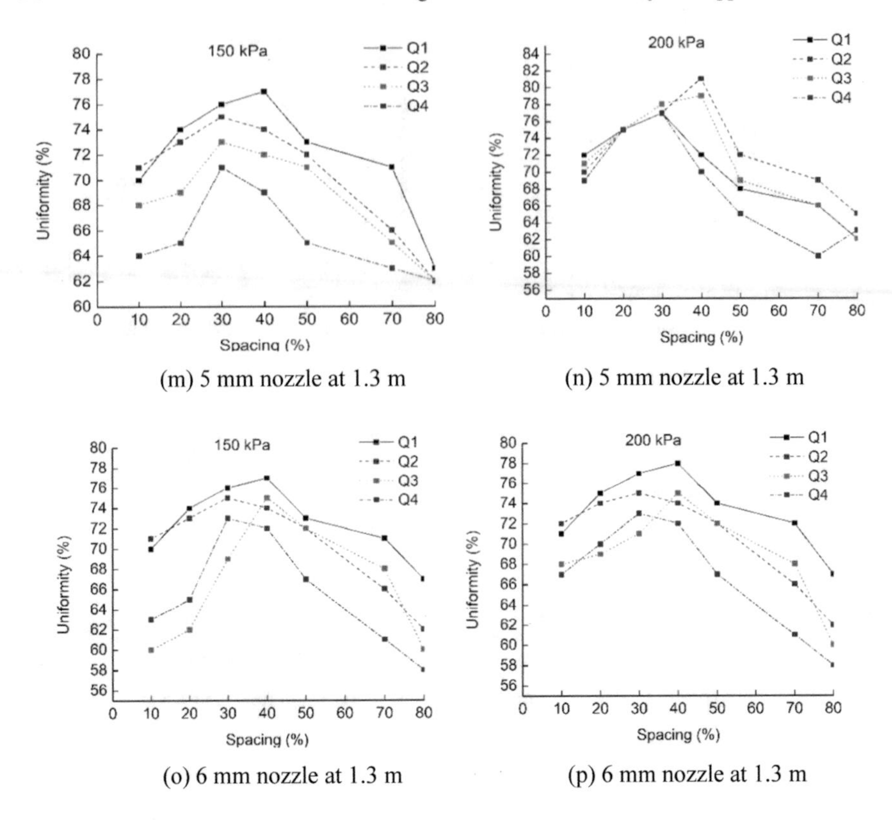

(m) 5 mm nozzle at 1.3 m

(n) 5 mm nozzle at 1.3 m

(o) 6 mm nozzle at 1.3 m

(p) 6 mm nozzle at 1.3 m

Fig. 4.7 (continued)

References

1. Duan FY, Liu JR, Fan YS, Chen Z, Han QB, Cao H (2017) Influential factor analysis of the spraying effect of light hose-fed traveling sprinkling system. J Drain Irrig Mach Eng. 35(6):541–546
2. Hills DJ, Barragan J (1998) Application uniformity for fixed and rotating spray plate sprinklers. Appl Eng Agric 14(1):33–36
3. Tang LD, Yuan SQ, Qiu ZP (2018) Development and research status of water turbine for hose reel irrigator. J Drain Irrig Mach Eng 36(10):963–968
4. Vories ED, Bernuth RD (1986) Single nozzle sprinkler performance in wind. Trans ASAE 296:1325–1330
5. Bahçeci T, Aydtn (2008) Determination of some performance parameters in set move sprinkler irrigation systems in Klzdtepe Plain of Mardin. J Fac Agric Harran Univ 32:435–449
6. Yuan SQ, Darko RO, Zhu XY, Liu JP, Tian K (2017) Optimization of movable irrigation system and performance assessment of distribution uniformity under varying conditions. Int J Agric Biol Eng 10(1):72–79
7. Zhu XY, Chikangaise P, Shi WD, Chen WH, Yuan SQ (2018). Review of intelligent sprinkler irrigation technologies for remote autonomous system. Int J Agric Biol Eng 11(1):23–30
8. Dogan E, Kirnak H, Dogan Z (2008) Effect of varying the distance of collectors below a sprinkler head and travel speed on measurements of mean water depth and uniformity for a linear move irrigation sprinkler system. Biosys Eng 99(2):190–195

9. Li J (2000) Sprinkler irrigation hydraulic performance and crop yield. Thesis report. Faculty of Agriculture, Kagawa University, Japan, pp 75–85
10. Xu ZD, Xiang QJ, Waqar AQ, Liu J (2018) Field combination experiment on impact sprinklers with aerating jet at low working pressure. J Drain Irrig Mach Eng 36(9):840–844
11. Hua L, Jiang Y, Li H, Zhou XY (2018) Hydraulic performance of low pressure sprinkler with special-shaped nozzles. J Drain Irrig Mach Eng 36(11):1109–1114
12. Hu G, Zhu XY, Yuan SQ, Zhang LG, Li YF (2019) Comparison of ranges of fluidic sprinkler predicted with BP and RBF neural network models. J Drain Irrig Mach Eng 37(3):263–269
13. Dwomoh FA, Yuan SQ, Li H (2013) Field performance characteristics of a fluidic sprinkler. Appl Eng Agric 29:529–536
14. Zhu X, Yuan S, Jiang J, Liu J, Liu X (2015) Comparison of fluidic and impact sprinklers based on hydraulic performance. Irrig Sci 33:367–374
15. Liu JP, Liu WZ, Bao Y, Zhang Q, Liu XF (2017) Drop size distribution experiments of gas-liquid two phase's fluidic sprinkler. J Drain Irrig Mach Eng 35:731–736
16. Liu JP, Liu XF, Zhu XY, Yuan SQ (2016) Droplet characterization of a complete fluidic sprinkler with different nozzle dimensions. Biosyst Eng 65(4):2-529
17. Xu SR, Wang XK, Xiao SQ, Fan ED (2019) Numerical simulation and optimization of structural parameters of the jet-pulse tee. J Drain Irrig Mach Eng 37(3):270-276
18. Zhu XY, Yuan SQ, Liu JP (2012) Effect of sprinkler head geometrical parameters on hydraulic performance of fluidic sprinkler. J Irrig Drain Eng 138(11):1019–1026
19. Zhu XY, Fordjour A, Yuan SQ, Dwomoh F, Ye DX (2018) Evaluation of hydraulic performance characteristics of a newly designed dynamic fluidic sprinkler. Water 10:1301. https://doi.org/10.3390/w10101301
20. Liu J et al. Droplet motion model and simulation of a complete Fluidic Sprinkler. Transactions of the ASABE, 61(4):1297-1306. https://doi.org/10.13031/trans.12639
21. Dukes MD (2006) Effect of wind speed and pressure on linear move irrigation system uniformity. Appl Eng Agric 22(4):541–548
22. Sourell H, Faci JM, Playan E (2003) Performance of rotating spray plate sprinklers indoor experiments. J Irrig Drain Eng 129(5):376–380
23. Xiang QJ, Xu ZD, Chen C (2018) Experiments on air and water suction capability of 30PY impact sprinkler. J Drain Irrig Mach Eng 36(1):82–87
24. Liu, JP, Yuan SQ, Li H, Zhu XY (2013) Numerical simulation and experimental study on a new type variable-rate fluidic sprinkler. J Agric Sci Technol 15(3):569-581
25. Dwomoh FA, Yuan SQ, Li H (2014) Sprinkler rotation and water application rate for the newly designed complete fluidic sprinkler and impact sprinkler. Int J Agric Biol Eng 7:38–46
26. Li J, Kawano H, Yu K (1994) Droplet size distributions from different shaped sprinkler nozzles. Trans ASAE 37:1871–1878
27. Keller J. Bliesner RD (1990) Sprinkler and trickle irrigation. An Avi Book, Van Nostrand Reinhold Pun, New York, p 651
28. Tarjuelo JM, Montero J, Carrion PA (1999) Irrigation uniformity with medium size sprinkler, part II: influence of wind and other factors on water distribution. Trans ASAE 42(3):677–689

Chapter 5
Comparative Evaluation of Hydraulic Performance of a Newly Design Dynamic Fluidic, Complete Fluidic, and D3000 Rotating Spray Sprinklers

Abstract This chapter presents the comparison of DFS, CFS and D3000 rotating spray sprinklers at different operating pressures. The following conclusions were made. The results showed that the discharge coefficient of the DFS and D3000 was slightly larger than that of the CFS. The water distribution profiles of the DFS, D3000, and CFS were parabola-shaped, ellipse-shaped, and doughnut-shaped, respectively. It was observed that CFS had the largest radius of throw under 200 kPa than DFS and D3000. The mean value of the coefficient of discharge was 0.83, 0.86 and 0.0.843, CFS, D 3000 and DFS, respectively. The comparison of velocity distribution showed that the maximum frequency value was obtained at velocities of 1 ms $^{-1}$ for each combination. Velocities for DFS and D3000 droplets were similar but not identical. Overall, CFS tends to give greater velocities than DFS or D3000. Individual spray sprinkler water distributions were mathematically overlapped to calculate the combined uniformity coefficient (CU). Maximal combined CUs of 81.83, 81.2, and 80.83% were found for the DFS, D3000, and CFS, respectively. Both the DFS and D3000 were found to have greater CU values than the CFS, which indicates that the DFS and D3000 provided a better water distribution pattern than the CFS at low pressure.

Keywords Uniformity · Droplet sizes · Velocity distribution

5.1 Introduction

More attention has been paid to the sustainability and efficiency of irrigation resources utilization, the use of low-pressure sprinkler irrigation is increasing all over the world, it is urgent to replace the existing high-pressure sprinkler irrigation with low-pressure sprinkler irrigation. The hydraulic performance of the commercial D3000 sprinkler, for example, is recognized as very good and is well-known to be low-pressure sprinklers. The dynamic fluidic sprinkler is comparatively cheaper, simple to construct and low consumption of energy. The complete fluidic sprinkler has numerous advantages that make it commonly used compared with other kinds of sprinklers, but its performance under low-pressure conditions has been recognized as unsatisfactory. Therefore, a comparative study on hydraulic performance for the

X. Zhu et al., *Dynamic Fluidic Sprinkler and Intelligent Sprinkler Irrigation Technologies*, Smart Agriculture 3, https://doi.org/10.1007/978-981-19-8319-1_5

rotation speed, spray range, uniformity, application rate, droplet sizes, and velocity distribution was performed. The basis for using these three sprinkler heads was to determine the performance quality of the newly designed dynamic fluidic sprinkler.

5.2 Structure and the Working Principle of Three Different Sprinkler Heads

The DFS and CFS were designed by the Research Centre of Fluid Machinery Engineering and Technology Jiangsu university china. The Nelson D 3000 spray head was from the Nelson irrigation Co.; Walla, Walla Washington USA. The main difference between the dynamic fluidic sprinkler (DFS) and complete fluidic sprinkler (CFS) is the working principle. The dynamic fluidic sprinkler receives an air signal from a signal tank, but the complete fluidic sprinkler obtains the signal from the fluidic component, found in the working area. In summary, the working theory of the (D3000) uses rotating fixed flat plates to break up the nozzle jet and create discrete streams of water leaving the plate edge. Table 5.1 presents corresponding operating pressures, nozzle sizes and flow rates for the three sprinkler heads (Fig. 5.1).

5.2.1 Experimental Setup and Procedure

The sprinkler heads were mounted on 2 m riser at 90° angle to the horizontal. A centrifugal pump was used to supply water from a constant level reservoir. Catch cans used in performing the experiments were cylindrical, 200 mm in diameter and

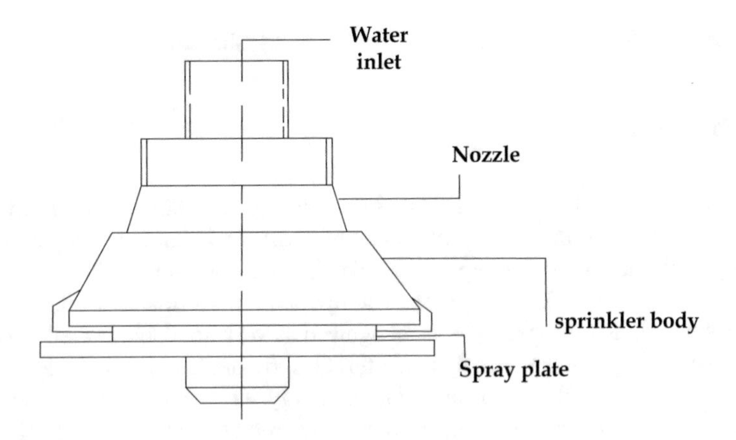

Fig. 5.1 Schematic, pictorial view D3000 sprinkler head

Table 5.1 Performance parameters of different types of sprinkler

Sprinkler type	Nozzle diameter (mm)	Pressure (kPa)	Flow rate (m^3/h)
DFS	5	100	0.76
		150	0.78
		200	0.80
D 3000	5	100	0.72
		150	0.74
		200	0.76
CFS	5	100	0.74
		150	0.77
		200	0.79

600 mm in height. The duration of each test lasted for an hour and working pressure was varied from 100 ~ 200 kPa. The water collected in each can was measured directly from a calibrated catch can. The application rate was calculated according to the water depth in each can. The diameter and velocity of the sprinklers were measured by laser precipitation monitor (LPM) from Thies Clima in Germany. The measurement of LPM is range from 0.125 mm to 8.0 mm. Droplet size measurements were divided into 0.1 mm increment (+0.05 mm) for analysis starting with 0.25 mm continuing to 7.95 mm. Measured drops less than 0.2 mm in diameter were discarded because they account for less than 0.01% of the total volume of the measured drops. The sprinkler nozzle was positioned 2 m above the laser beam of the LPM. The measurements of droplet size were collected at the middle (6 m) and end (12 m) of the radial distance from the nozzle (Fig. 5.2).

5.2.2 *Calculation of Combined CUs, Droplet Sizes, and Velocities*

The uniformity of water application rate was calculated using the Christiansen coefficient of uniformity (CU) in Eq. 1.5. Uniformities for the sprinklers were simulated using Matlab software under different operating pressures. Cubic spline interpolation was used to analyze the water distribution data of a single nozzle and the measured data was converted to grid type. Then, the superposition method was adopted to calculate the combined uniformity coefficients for the overlapped sprinklers and the optimal combined water distribution maps were drawn. To avoid the phenomenon of missed spraying, the value of the combination spacing varied from R to 1.8R. Through the statistical analysis of the droplet data set, the parameters of centrality and dispersion of droplet size were obtained. Droplet size parameters used in this study include average volume diameter (Dv, mm) and arithmetic mean diameter

Fig. 5.2 Schematic diagram of the experimental apparatus

(d, mm) were calculated using Eqs. 3.8 and 3.9, respectively. The discharge coefficients of each nozzle were determined for the observed pressure-discharge data using Eq. (1.1).

5.3 Results and Discussion

5.3.1 Comparison of a Radius of Throw and Coefficient of Discharge at Different Operating Pressures

Table 5.2 presents the summarized of discharge and spray range of the three sprinkler heads. The average discharge and standard deviation for all the sprinkler heads are shown in Table 5.2. The coefficient of discharge for CFS sprinkler ranged from 0.72 ~ 0.92 with an average value of 0.83, while that from the DFS and D3000 sprinkler were from 0.74 ~ 0.93 with an average of 0.86 and 0.74 ~ 0.93 with an average of 0.843, respectively. The discharge coefficients for all the sprinkler heads changed slightly when the working pressure was increased. However, the discharge coefficient of DFS was higher than the rest of the sprinklers, which means that the DFS sprinkler had the advantages of higher irrigation intensities. These can be attributed to fewer restrictions within the inner flow movement. The new sprinkler (DFS), had the advantages of a larger discharge coefficient. These can be attributed to fewer restrictions within the inner flow movement. The comparison of the radius of throw for the three sprinklers under the same working conditions demonstrated

Table 5.2 Radius of throw and coefficient of discharge of three sprinkler types at different operating pressures

Radius of throw (m)							Discharge coefficient				
Sprinkler type	Nozzle size (mm)	Pressure (kPa)	150	200	250	Standard deviation	Pressure (kPa)	150	200	250	Standard deviation
CFS	5		8.2	10.1	11.5	1.66		0.72	0.90	0.92	0.096
DFS	5		12.2	10.5	10.2	1.07		0.93	0.91	0.74	0.10
D3000	5		11.5	11.7	11.5	0.115		0.74	0.86	0.93	0.09

that DFS gave a higher distance of throw. It was observed that as pressure increased the distance of throw also increased for all the sprinkler heads. The maximum spray range from the DFS, D3000, and CFS was 12.3, 11.3, 11.5 m and the standard deviation for three sprinklers was 1.07, 0.115 and 1.66, respectively. It was found that the spray ranges from the D3000 were smaller and this could be attributed to the degree of interruption causing a small reduction in the distance traveled by the water jet.

5.3.2 Relationship Between Rotation Speed for Three Different Sprinkler Heads

Figure 5.3 presents the results of quadrant completion time under different operating pressures for all the sprinkler heads. Generally, for all the sprinkler heads, increasing

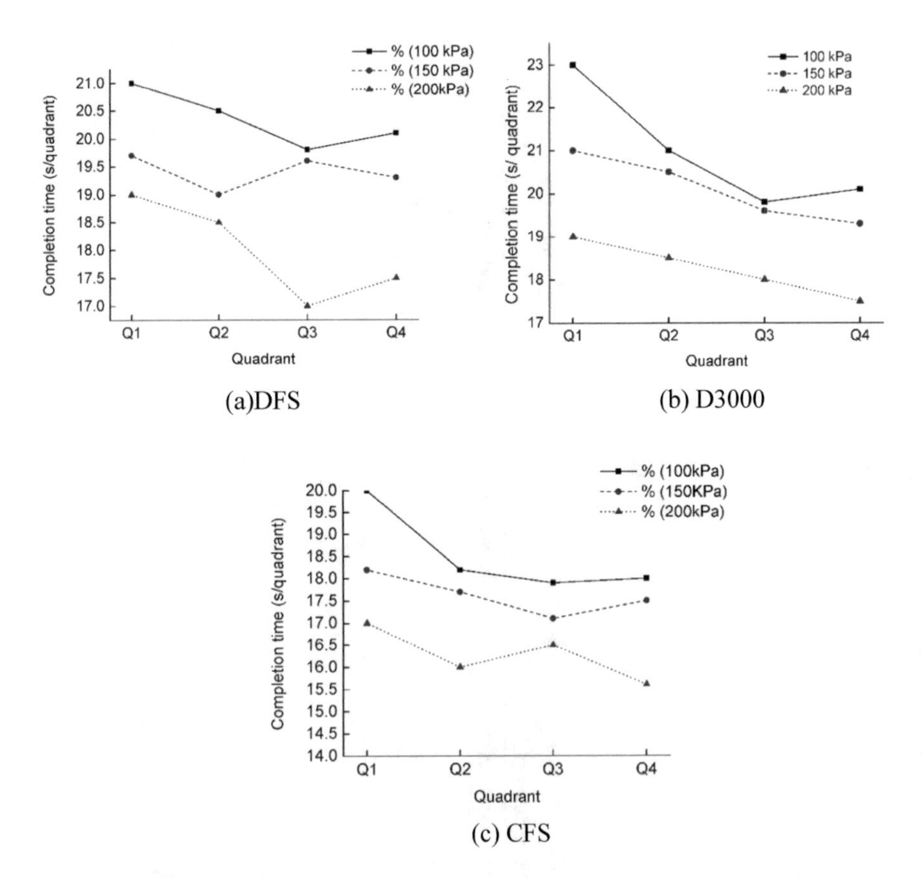

Fig. 5.3 Quadrant completion time under different operating pressures for both sprinkler heads

the pressure resulted in a decrease in the speed of rotation. It was found that variations in completion time through the quadrant were higher for CFS. This is in agreement with the findings. The variations in quadrant completion times were 20, 18.8, 17.9, and 18 s for Q1, Q2, Q3, and Q4, respectively as shown in Fig. 6.4. It was found that the differences in quadrant completion times were significant especially with 150 kPa, this clearly shows that CFS sprinkler performed poorly under low-pressure conditions. This occurrence is due to the stepwise rotation of the CFS sprinkler which accounts for the fast wall attachment principle that takes place within the fluidic element. This observation is in quite an agreement with the observation of [1–4]. The variations in quadrant completion time at different pressures showed that DFS and D3000 were lower compared with CFS, especially under low-pressure conditions. The variations in quadrant completion times were 21, 20.6, 19.9, and 20.3 s for Q1, Q2, Q3, and Q4, respectively as shown in Fig. 6.4 for DFS. The possible reason could be that pressure variations were quite uniform. The results of the quadrant completion times showed that under low-pressure conditions, DFS produced optimum rotation stability compared with CFS.

5.3.3 Comparison of Water Distribution Profiles

A comparison of water distribution across the radial lines for the DFS, CFS and D3000 at 100, 150 and 200 kPa are shown in Fig. 5.4. For all the sprinklers, the water application rate was higher close to the sprinkler and reduced linearly as the distance from away from the sprinkle. Generally, at higher operating pressures water application intensities were smaller and more uniformly distributed. For the various operating pressures, the average values of the DFS application rate changed from 3.9 to 7.6 mm h^{-1}. The parameters combination of 100 kPa × 7.0 mm h^{-1} × 8 m; 150 kPa × 6.1 mm h^{-1} × 10 m and 200 kPa × 6.32 mm h^{-1} × 7 m recorded the maximum application rate. The application rate decreased suddenly as the distance increased from the sprinkler until it reached a minimal level. Starting from this distance, the application rate decreased gradually to reach the minima. For 100 kPa at 10 m, 150 kPa at 10 m and 200 kPa at 11 m, the minimum values were 6.2, 5.8, and 3.9 mm h^{-1}, respectively. A similar trend was observed for D3000 and CFS. The water application rate increased to a maximum value as the distance from the sprinkler increased and then decreased gradually. The application rates of D3000 changed from 0.42 to 7.3 mm h^{-1}. The maximum value of application rate recorded for the three evaluated pressures was (6.00 mm h^{-1} at 7 m for 100 kPa, 6.3 mm h^{-1} at 5 m for 150 kPa and 6.7 mm h^{-1} at 7 m for 200 kPa). Indicating that the average application rate of each sprinkler is inversely related to sprinkler wetted radius since the flow rate of the sprinklers was nearly the same and sprinkler spacing along the lateral was equal. The application rate decreased suddenly as the distance increased from the sprinkler until it reached minimal. A similar trend was also observed for CFS, application rates for CFS varied from 1.96 to 6.7 mm h^{-1}. The maximum value of application rate recorded for the three evaluated pressures was (6.3 mm h^{-1} at 7 m

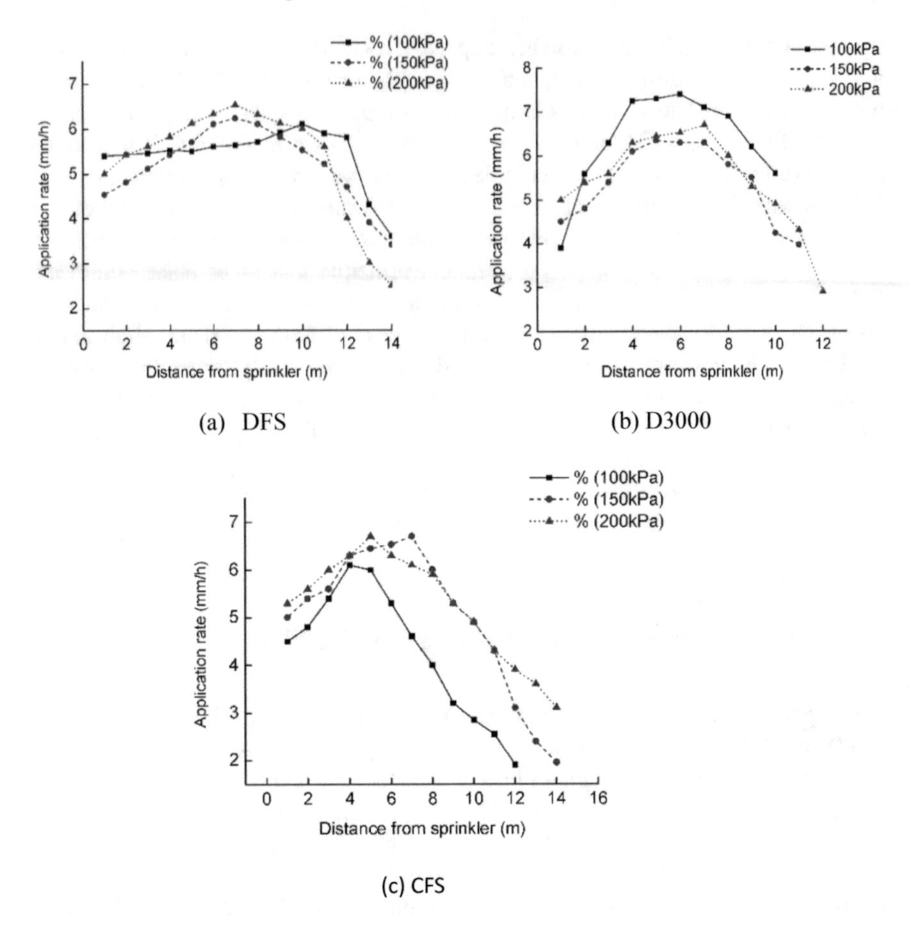

Fig. 5.4 Water application profiles for different operating pressures

for 100 kPa, 6.7 mm h^{-1} at 6.7 m for 150 kPa and 7 mm h^{-1} at 5 m for 200 kPa,). This quite agrees with finding [5]. Starting from this distance, the application rate decreased sharply to reach the minima. For 100 kPa at 10 m, 150 kPa at 10 m and 200 kPa at 11 m, the minimum values were 5.5, 3.2, and 4.2 mm h^{-1}, respectively. The water distribution of three sprinklers under the same conditions was compared. The results showed that the average maximum water distribution of CFS was lower than that of DFS and D3000. These differences can be attributed to two factors. Firstly, under the same operating pressure, the flow rate of DFS and D3000 is much higher than that of CFS. Secondly, the instantaneous water utilization rate of CFS changes several times from the minimum to the maximum and then recovers to zero. Under the conditions of DFS and D3000, the instantaneous sprinkler water application rate does not change in a wide range, which leads to the higher average maximum water application rate of these sprinklers. Similar to those obtained by [6] who reported that increasing nozzle diameter usually increases the average application rate since the sprinkler discharge tends to increase more rapidly than the wetted area. Among the pressures, 150 kPa performed better than 200 kPa for all the sprinklers. For all the sprinklers it was observed that as operating pressure was increased, the application rates increased until they reached the maximum when they started to decrease.

5.3.4 Comparison of the Computed Uniformity Coefficient

Figure 5.5 shows the relationships between the combined CU and spacing along the main axis for all the sprinkler types. In the study, the square spacing for lateral radius times of 1.0, 1.1, 1.2, 1.3, 1.4, 1.5, 1.6, 1.7, and 1.8, was selected for each of the sprinklers. Different operating pressures were used for the calculation for the

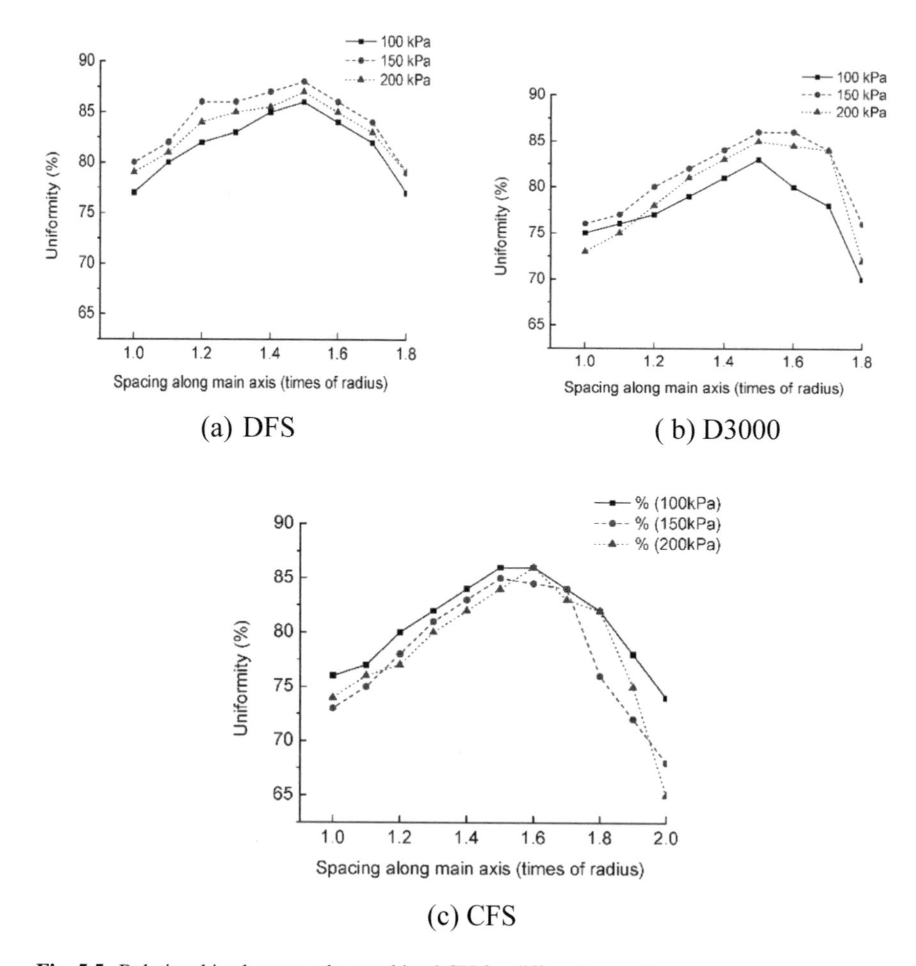

Fig. 5.5 Relationships between the combined CU for different operating pressures

sprinklers and these include 100, 150 and 200 kPa. From Fig. 6.6 it can be observed that the CU value increased as the distance between the sprinklers increased until it reached a maximum and reduced for all the pressures for all the sprinkler types. The range of combined CUs value recorded at different pressures for DFS was as follows: 75% at a spacing of 1 (time of wetted radius) to 65% at 1.8 times (100 kPa), 76% at 1 time to 70% at 1.8 times (150 kPa). The average combined uniformity values obtained for DFS at different operating pressures were 76.2, 81 and 79.2% 100, 150 and 200 kPa, respectively. The highest CU recorded at 1.5 times spacing and uniformity increased with spacing 1 to 1.5, ranging from 76 to 86% with an average 81.83%. Thereafter, the uniformity decreased with spacing. From 1.6 to 1.8 times spacing, the CU ranged from 74 to 86% with an average of 78.94% at an operating pressure of 150 kPa. The range of combined Cu value recorded at different pressures for D 3000 was as follows: 75% at a spacing of 1 (time of wetted radius) to 65% at 1.8 times (100 kPa), 76% at 1 time to 70% at 1.8 times (150 kPa). The average combined uniformity values obtained for D 3000 at different operating pressures were 76, 80.2 and 80% 100, 150 and 200 kPa, respectively. The highest CU recorded at 1.5 times spacing and uniformity increased with spacing 1 to 1.5, ranging from 76 to 86% with an average 80.83%. Thereafter, the uniformity decreased with spacing. From 1.6 to 2 times spacing, the CU ranged from 74 to 86% with an average of 78.94% at an operating pressure of 150 kPa. For the different pressures, the uniformities increased with spacing from 1 to 1.8 times with an average of 81.2, 78.5 and 77%, respectively. Similarly, the computed CUs from type CFS increased from 69% at R to 83% at 1.6R (100 kPa), 69% at R to 84% at 1.6R (150 kPa) and 72% at R to 85% at 1.6R (200 kPa) with average uniformities of 75.6, 76.29and 77.04%, respectively. The highest CU recorded at 1.5 times spacing and uniformity increased with spacing 1 to 1.5, ranging from 76 to 86% with an average value of 80.83%. Thereafter, the uniformity decreased with spacing. These occurrences agree with published results [7]. The comparison of the CU from the three types of sprinkler heads reveals that there was little difference between the calculated CU values for the DFS and D3000. Both DFS and D3000 gave higher CU values than the CFS for a given pressure and sprinkler spacing, especially when operated at low pressure. This indicates that the DFS and D3000 provided a better water distribution pattern than the CFS at low pressure. One possible explanation for this could be that for the two-phase DFS, the internal turbulent flow was less uniform from the nozzle outlet, and more water was applied near the sprinkler, resulting in a higher combined CU. The reduction in the water distribution uniformity indicated that the CFS was not too perfect in delivery of irrigation water, hence will affect crops grown under such irrigation as they will not receive more water thereby increasing the energy required as well as the cost involved.

Table 5.3 Results of spray distribution in the middle and end

	Operating pressures (kPa)			
Sprinkler type	Measuring position	100	150	200
CFS	Middle	3.17	3.1	3.91
	End	5.14	4.78	4.2
DFS	Middle	3.3	3	2.7
	End	4.2	3.5	3.2
D3000	Middle	3.6	3.3	2.95
	End	4.7	4.2	3.95

5.3.5 Spray Distributions in the Middle and End of the Range

Spray distribution in the middle and end of the range was obtained by Laser Precipitation Monitor for the sprinkler types. Table 5.3 presents the details results of spray distribution in the middle and end. From the table, it can be seen that droplet sizes obtained in the middle and end for all the sprinkler heads decreased with increasing operating pressure. However, the droplet diameters in the middle positions of the DFS and D3000 sprinkler under different pressures were much smaller than the CFS. This explanation could be that the internal flow was less uniform from the nozzle outlet, and more water was applied near the sprinkler, resulting in a large droplet size at middle. Moreover, droplet size produced from both sprinkler heads in the end positions was smaller and similar in sizes, which means all the sprinklers were effective in reducing the end droplet diameters of spray in the end position, a similar trend was also observed for the radial water distributions given in Fig. 5.5. This could be necessary to minimize the effect of runoff and reduce erosion in the soil.

5.3.6 Droplet Size Distribution

A comparison of droplet size characterization of DFS, CFS and D3000 was determined under indoor conditions. Figure 5.6 presents droplet sizes form the three sprinkler types at different operating pressures. Form the three sprinkler heads, larger droplet sizes were produced at low pressure and as pressure was increased small droplet sizes were recorded. From the study, it was observed that as the distance to sprinkle increases, the frequency of large drops increases. The cumulative frequency graph shows that small drops (<2 mm of diameter) exceeded 80% frequency for the distance below 10 m at the large distances, 4 and 5, the curves were less steep in frequency, indicating that the distribution of diameter was well graded. Similar results have been obtained by researchers upon using different types of sprinklers [8–10]. The mean DFS droplet diameters ranged from 0 to 3.2 mm. The cumulative frequencies under 1 mm, 75, 87 and 85%, under 2 mm of 97, 96 and 91% under 3 mm

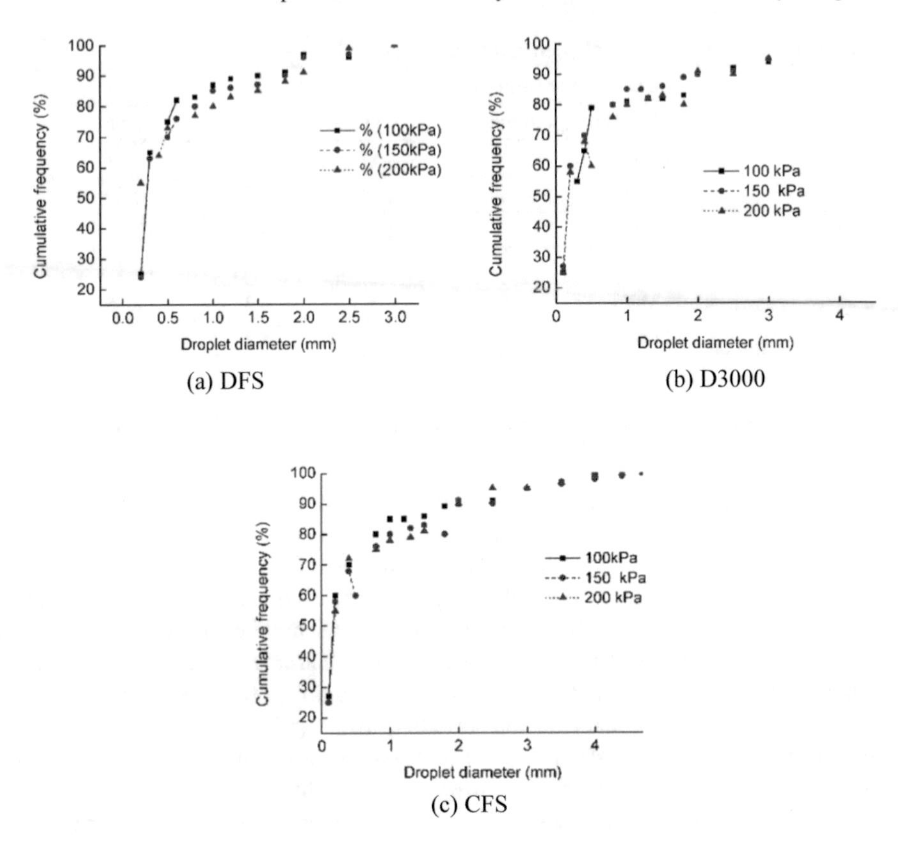

Fig. 5.6 Cumulative droplet diameter frequency

of 99,98 and 91% at pressures of 100,150, and 200 kPa respectively. The average D 3000 droplet diameters ranged from 0 to 4 mm. The cumulative frequencies under 1 mm, 81, 85, and 99%, under 2 mm of 90, 90 and 91% under 3 mm of 94, 95, and 95% under mm of 99,98and 99% at pressures of 100,150 and 200 kPa respectively. The mean CFS droplet diameters varied from 0 to 4.7 mm. As a consequence, droplets under 1 mm had cumulative frequencies of 84, 72, 73 and 70%, under 2 mm of 91, 94, 94 and 94%, under 3 mm of 97, 97, 96 and 97 for the pressure of 100, 150, and 200 kPa, respectively. The comparison of droplet size distribution from the three types of sprinkler shows that DFS and D3000 were similar to each other, DFS had the narrowest droplet size range and the smallest maximum droplet diameter (about 3.2 mm), and CFS had the widest droplet size range with a maximum value of about 4.7 mm and a clear trend were observed between diameter and cumulative frequency, small drops are observed in the vertical trajectory while larger drops are observed when the trajectory has a relevant horizontal component. CFS also had an approximately 0.5 mm larger droplet diameter than DFS and D3000 (Table 5.4).

Table 5.4 Summary of droplet size distribution (mm) for 10, 25, 50, 75 and 90% (d10, d25, d50, d75 and d90, respectively) of different types of nozzle

Sprinkler type	Pressure (kPa)	d_{10}	d_{25}	d_{50}	d_{75}	d_{90}	Standard deviation
DFS	100	0.1	0.2	0.3	1.3	2.2	0.9
	150	0.2	0.3	0.4	0.9	2	0.76
	200	0.2	0.3	0.6	0.9	1.8	0.80
D3000	100	0.05	0.17	0.31	1.07	1.53	0.64
	150	0.07	0.14	0.25	1.07	1.82	0.76
	200	0.06	0.13	0.23	1.28	1.84	0.80
CFS	100	0.07	0.48	0.51	1.2	1.53	0.61
	150	0.08	0.31	0.35	1.32	1.40	0.61
	200	0.07	0.28	0.30	1.40	1.50	0.68

d10 = represents 10% of the cumulative droplet frequency; d25 = represents 25% of the cumulative droplet frequency; d50 = represents the mean cumulative droplet frequency; d75 = represents 75% of the cumulative droplet frequency; d90 = represents 90% of the cumulative droplet frequency

5.3.7 *Droplet Characterization Statistic*

Table 5.5 shows basic drop characterization statistics and velocity obtained from the three sprinkler heads at different operating pressures. As expected, the drop diameter increased with distance. This can be seen for D_v and D_{50}. The arithmetic drop diameter does not reveal this trend since it assigns the same averaging weight to drops at different diameters. The following were observed for all the sprinkler heads, the standard deviation of drop diameter increased with distance from the sprinkler, and so did the coefficient of variation, which attained its maximum value (30) for all the sprinklers. However, the maximum observed drop diameter increased with distance, indicating that large drops travel distance proportional to their diameter. For DFS, the standard deviation of droplet diameter ranged from 0.94 to 1.01, with an average of 0.98; the coefficient of diameter variation ranged from 113 to 130, with a mean of 122.2. For D 3000, the standard deviation of droplet diameter ranged from 0.92 to 1.0, with an average of 0.96; the coefficient of diameter variation ranged from 110 to 130, with a mean of 121.7. For CFS, a standard deviation of droplet diameter ranged from 0.94 to 1.01, with an average of 0.98; the coefficient of diameter variation ranged from 113 to 130, with a mean of 122.2. Regarding drop velocity, the coefficient of variation and standard deviation increased with distance from the sprinkler for both sprinkler heads. For DFS, the standard deviation of velocity ranged from 1.32 to 1.38, with an average of 1.35; the coefficient of velocity variation ranged from 59 to 76, with a mean of 66.33. For DFS, a standard deviation of velocity ranged from 1.29 to 1.47, with an average of 1.38; the coefficient of velocity variation ranged from 71 to 79, with a mean of 72.3.

Table 5.5 Statistical analysis of operating pressure and flow velocity effects on droplet diameter

Sprinkler type		DFS			D3000			CFS		
Pressure (kPa)		100	150	200	100	150	200	100	150	200
Diameter (mm)	$\bar{d}$	0.89	0.76	0.74	0.74	0.68	0.65	0.69	0.72	0.73
	d_v	2.78	2.74	2.16	2.7	2.12	1.43	2.5	2.61	1.57
	d_{50}	0.22	0.21	0.2	0.34	0.25	0.2	0.21	0.25	0.28
	SDD	1.0	0.99	0.92	1	0.99	0.92	0.88	0.96	0.94
	CVD	112	128	123	115	125	129	112	129	136
Velocity (ms^{-1})	V	1.68	1.97	2.14	1.87	1.92	1.48	1.59	1.95	1.46
	SDV	1.38	1.32	1.34	1.47	1.29	1.43	1.36	1.33	1.39
	CDV	76	64	59	79	67	71	72	68	73

$\bar{d}$ = arithmetic mean droplet; dv = the volume weighted average droplet diameter; SD_D = the standard deviation; CV_D = is the coefficient of variation

5.3.8 *Droplet Velocity Distribution*

Figure 5.7 presents the relationship between frequency distribution and droplet velocity form from the sprinkler heads at different operating pressures. The following observations were made; Mean velocities ranged from 0 to 5.7 ms^{-1} were recorded under DFS. Frequency of 23, 20 and 21.5% were observed at 1 ms^{-1}, under 3 ms^{-1} of 4.7, 6 and 7%, under 4 ms^{-1} of 3.8, 2.5 and 3.2%, under 5 ms^{-1} of 3.8, 20.5, and 3.1% for a pressure of 100, 150, and 200 kPa, respectively. Mean velocities ranged from 0 to 6.3 ms^{-1} were also obtained under D3000. Frequency of 27.5, 22.5 and 20% was observed at 1 ms^{-1}, under 3 ms^{-1} of 10, 8 and 7.3%, under 4 ms^{-1} of 4.8, 3.7 and 4.6% under 5 ms^{-1} of 4.5, 3.1 and 4.4% for a pressure of 100, 150, and 200 kPa, respectively. The mean CFS droplet velocities ranged from 0 to 6.7 ms^{-1}. Droplets under 1 ms^{-1} had a frequency of 23%, 17%, and and17%, under 3 ms^{-1} of 5%, 6%, and 6% and, under 5 ms^{-1} of 3%, 3%, and 2% for a pressure of 100, 150, and 200 kPa, respectively. This comparison shows that the maximum frequency value was obtained at velocities of 1 ms^{-1} for each combination. This comparison shows that the maximum frequency value was obtained at velocities of 1 ms^{-1} for each combination. Velocities for DFS and D3000 droplets were similar but not identical. Overall, CFS tends to give greater velocities than DFS or D3000. Also, the operating pressure and spaying distance from the sprinkler to target surfaces significantly affect droplet diameter.

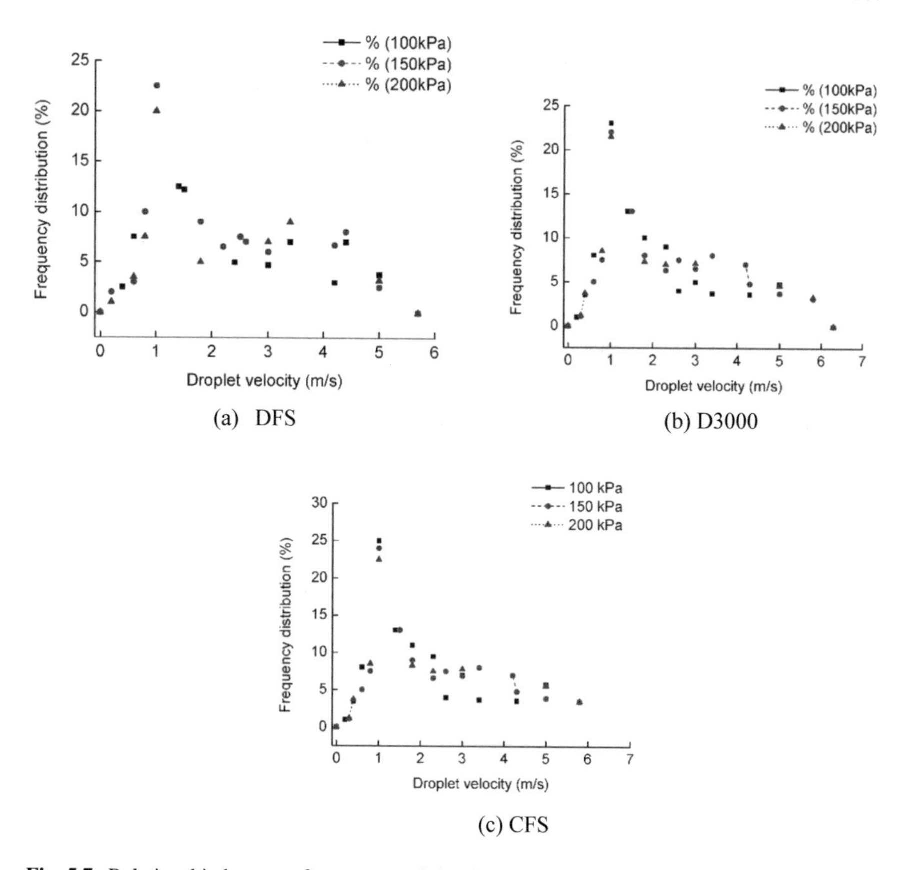

Fig. 5.7 Relationship between frequency and droplet velocity for both sprinkler heads at different operating pressures

References

1. Salles C, Poesen J, Borselli L (2012) Measurement of simulated drop size distribution with an Optical spectro pluviometer sample size concentration [J]. Earth surface processes landforms, 24(6):5-45
2. Liu J, Zhu X, Yuan S, et al. (2019) Modeling the application depth and water distribution of a linearly moving irrigation system [J]. Water, 10, 1301
3. Tang L D, Yuan S Q, Qiu Z P (2018) Development and research status of water turbine for hosereel irrigator [J]. J Drainage & Irrig Mach Eng, 36(10):963-968
4. Zhu XY, Chikangaise P, Shi W D, et al. (2018) Review of intelligent sprinkler irrigation technologies for remote autonomous system. Int J Agric & Biol Eng 11(1):23–30
5. Friso D, Bortolini L (2010) Calculation of sprinkler droplet-size spectrum from water distribution radial curve [J]. Int J Energy Technol 24:1–11

6. Montazar A, Moridnejad M (2008) Influence of wind and bed slope on water and soil moisture distribution in solid-set sprinkler systems[J]. Irrig Drain 57:175–185
7. Andales AA, Bauder TA, Arabi M A mobile irrigation water management system using a collaborative GIS and weather station networks [M]. In: Ahuja LR, Ma L, Lascano R (eds)
8. Zhang H, Oweis T (1999) Water–yield relations and optimal irrigation scheduling of wheat in the mediterranean region[J]. Agric Water Manag 38:195–211
9. Kundu DK, Neue H, Singh R (1998) Comparative effects of flooding and sprinkler irrigation on growth and mineral composition of rice in an Alfisol [C]. In: Proceedings of the national seminar on micro- irrigation research in India: status and perspective for the 21st Century. Bhubaneswar
10. Wrachien D, Lnrenzini G (2006) Modeling jet flow and losses in sprinkler irrigation: overview and perspective of a new approach [J]. Biosys Eng 94(2):297–309

Chapter 6
Modelling of Water Drop Movement and Distribution in No Wind and Windy Conditions for Different Nozzle Sizes

Abstract A numerical model was developed to determine the water drop movement and mean droplet size diameter at any distance from a sprinkler as a function of nozzle size and pressure. Droplet size data from 4, 5, 6, and 7 mm nozzle sizes verified the model. Data for model prediction were generated throughout lab experiments. The results demonstrated that the correlation between the observed and predicted droplet size diameter values for all the nozzle sizes and pressures is quite good. Nozzle size and pressure had a major influence on droplet size. Higher pressure produced smaller droplets over the entire application profile. The wetted distance downwind from the sprinkler increased as wind velocity increased, for example at a constant working pressure of 300 kPa, at wind speeds of 3.5 m/s and 4.5 m/s, 20% and 32% of the total volume exceeded the wet radius respectively. Larger droplets (3.9–4.5 mm), accounting for 3.6% and 6.3% of the total number of distributed droplets, respectively. The model can also predict the droplet size distribution at any wind direction overall the irrigated pattern.

Keywords Sprinkler irrigation · Nozzle size · Droplet size · Modeling

6.1 Introduction

Irrigation sprinklers deliver many water drops of different sizes, which have the characteristics of sprinkler nozzle and pressure configuration. The performance of sprinklers is usually classified as overlapping uniformity and droplet size distribution [1–4]. It is attributed to the physical characteristics of the sprinkler, nozzle configuration, working pressure, sprinkler spacing, and environmental conditions (wind speed and direction). In other words, the hydraulic performance of the sprinkler is a function of its physical characteristics, geometric parameters, and environmental conditions [5, 6, 7]. Therefore, the different types and sizes of sprinklers have different hydraulic performance characteristics. Working pressure and nozzle characteristics (nozzle opening size, shape, and angle) are the main factors controlling sprinkler performance [8, 9]. A study conducted by [10] demonstrated that the droplet size distribution from agricultural sprinklers showed that decreasing droplet size with increasing relative velocity of the water to the air, and Ref. [11] reported that nozzle

X. Zhu et al., *Dynamic Fluidic Sprinkler and Intelligent Sprinkler Irrigation Technologies*, Smart Agriculture 3, https://doi.org/10.1007/978-981-19-8319-1_6

pressure had a major influence on droplet size. It was established that a higher pressure produced smaller droplets over the application profile. The authors in [12] found that for both circular and square nozzles, increasing pressure decreased droplet size in overall droplet spectra, and for a given pressure, changing nozzle shape from circular to square also decreased droplet size.

The droplet size distribution varies with distance from the sprinkler [13–16]. Knowledge of droplet size distribution is important because they determine the effect of droplets from sprinklers on wind, evaporation, and impact on the soil surface [17, 18]. Manufacturers are interested in knowing the size, percentage, or volume of droplets and where they are deposited to compare products, evaluate designs, and predict the effects of operating conditions such as pressure. Over the years, several simulation studies have been conducted to simulate various aspects of the impact of wind on water droplets in sprinklers [19–24]. Many factors affect the trajectory and loss of water droplets, which complicates the overall description and estimation of water droplet drift. The authors, in [25–27], in 1995, studied the movement of water droplets in the air mainly affected by drag and gravity. The authors in Ref. [10], in 1989, determined the volume average droplet diameter at any distance from the nozzle according to the nozzle size and pressure. The model was validated by 4.0 mm, 3.2 mm circular, and 3.5 mm square nozzles. The authors in Ref. [28] studied the effect of wind and reported that the wind at right angles to the wind has lengthened the model. The authors confirmed that the size distribution of water droplets changes with the distance from the nozzle. However, it is important to understand the size distribution of water droplets, because they determine the effect of water droplets on wind, evaporation, and response to soil surface effects. The existing complete fluidic sprinkler is known for its rotation problems, particularly when operated at low-pressure conditions. Therefore, the fluidic component and nozzle were optimized, leading to the development of a new type of sprinkler called the dynamic fluidic sprinkler (DFS). In the current studies, no work has been conducted on the droplet size traveling distance of the (DFS). Therefore, it is very important to study the distribution of droplet size and the traveling distance of the (DFS), which is of great theoretical value and practical significance. These variables were measured using 2D-Video distrometer. The objective of this study was to develop a model for droplet size traveling distance and to verify the accuracy of the results through experimental data.

6.2 Materials and Method

6.2.1 Boundary Condition

The ballistic approach was adopted to model the drop trajectory until reaching the ground surface. The following assumptions were made; the movement of the drop is influenced by; (a) its initial velocity vector; (b) gravity acting in a vertical direction

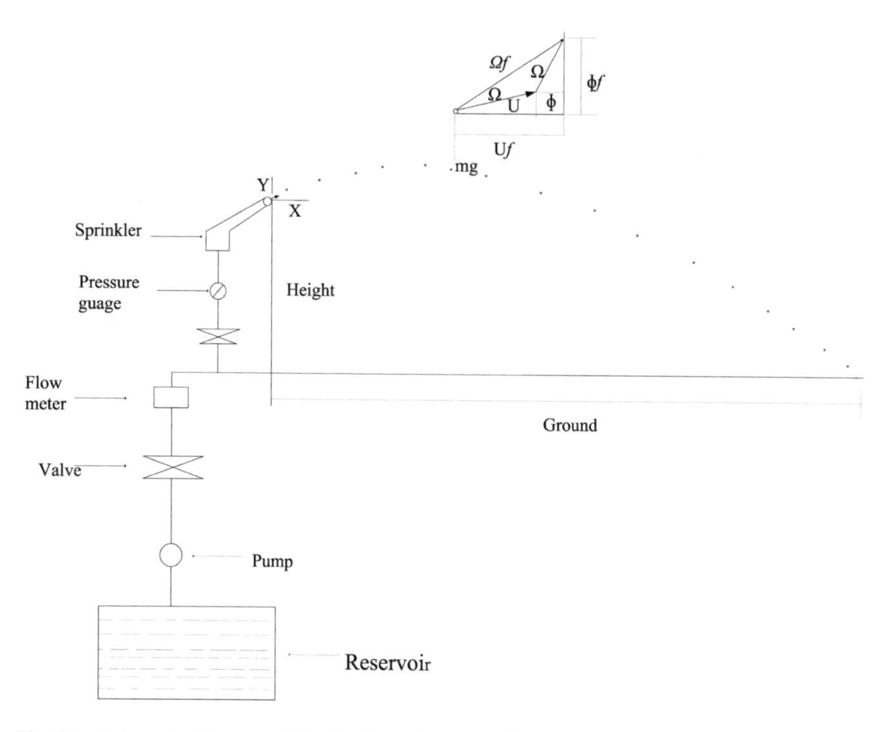

Fig. 6.1 Schematic diagram of the droplet trajectory under no wind conditions

(c) the wind vector acting in the horizontal plane; and (d) the resistance force applied in the direction opposite to the relative movement of the drop in the air. Due to the complexity of the spray jet process, the model considers the following aspects: (1) The jet is disintegrated at the nozzle exit into individual drops with different diameters moving independently in the air; (2) the drag coefficient is independent of the sprinkler height over the ground surface; and (3) different sized drops fall at a different distance. Droplet travel distance under no wind condition is undisturbed, and thus, a characteristic of the droplet size for a given configuration. The height of the sprinkler is 1.9 m, droplet diameter range considered; 0 < droplet diameter (mm) < 6; wind speed; 0, 2.5, 3.5 and 4.5 (m/s), operating pressure, 150, 200, 250 and 300 kPa. Nozzle sizes 4, 5, 6 and 7 mm.

6.2.2 *Model of Droplet Motion*

Figure 6.1 presents a schematic diagram of the droplet trajectory under no wind conditions. The effect of wind drift of sprinkler spray on drops distribution of a single sprinkler has been analyzed. The water jet ejected from a sprinkler nozzle is assumed to be a flux of spherical water drops having various drops diameter.

Therefore, the sprinkler discharge profile is determined by the trajectories of water drops and its volumetric distribution. Several models have been developed by considering a sprinkler as a device emitting numerous droplets of diameter as functions of their traveled distances [18, 29, 30]. The ballistic theory, equations of motion for discharged water drops were adopted and are expressed as:

$$\frac{d^2 k}{d t^2} = -\frac{3}{4}\frac{\rho_a}{\rho_x}\lambda_q\left(\mu_f - \mu\right) V_r \tag{6.1}$$

$$\frac{d^2 l}{d t^2} = -\frac{3}{4}\frac{\rho_a}{\rho_x}\lambda_q\left(\Phi_f - \Phi\right) V_r \tag{6.2}$$

$$\frac{d^2 p}{d t^2} = -\frac{3}{4}\frac{\rho_a}{\rho_x}\lambda_q\left(\beta_f\right) V_r - g \tag{6.3}$$

where *k* and *l* are the $\updownarrow_q$ distances in the horizontal and vertical directions (m), ρ_a, ρ_x is the density ratio of air and water, respectively; *t* is time (s) and g is the acceleration due to gravity. $\updownarrow_q$ is the air drag coefficient of the droplet moving at the speed, V_r.

$$V_r = \sqrt{\left[\left(\mu_f - \mu\right)^2 + \left(\Phi_f - \Phi\right)^2 + {\beta^2}_f\right]} \tag{6.4}$$

While μ and σ are the horizontal and vertical components of the droplet velocity, respectively; μ_f, σ_f and β_f are the x, y, and z components of the wind velocity, respectively. Given that the logarithmic profile of wind speed is generally considered to be a reliable estimator of the actual field condition, the absolute wind speeds were calculated for all the conditions is:

$$W_r = W_m = \frac{In\left[{(n - B)}/{n_o}\right]}{In\left[{(n_r - B)}/{n_o}\right]} \tag{6.5}$$

W_m = air velocity (m/s) measured at reference height, n_r (m) above ground. B and n_0 are roughness height (m) and roughness parameter (m) respectively, both are functions of crop height (*h*), given by:

$$\log d = 0.997 \log h - 0.1536 \text{ and } \log n_o = 0.997 \log h - 0.883 \tag{6.6}$$

The fourth-order Runge–Kutta numerical integration techniques were used to solve Eqs. 6.1–6.3 for droplet movement.

6.2.3 Empirical Model of the Drag Coefficient

The drag force acting on the trajectories and contact points of drops discharged from the sprinkler nozzle under the same pressure was determined by using Eqs. 6.5 and 6.6:

$$\lambda_q = \left[\left({}^{a}/_{R_e}\right)^{c} + b^{c}\right]^{1/2} \tag{6.7}$$

$$\upsilon = \lambda_q \,(2gH)^{1/2} \tag{6.8}$$

where, $a = 24$; $b = 0.32$; $c = 0.52$; D the droplet diameter (m) and υ, the kinematic viscosity of air ($m^2\ s^{-1}$). The adopted relationship compares very well with the well-known set of relations by [29]. The model applies not only to the turbulent-flow regime, but also to the Stokes regime. However, it shows some deviation from experimental data for $Re > 10^4$. The velocity of the sprinkler jet exiting from the nozzle was calculated as follows:

$$V = CD(2gH)^{0.5} \tag{6.9}$$

where H (m) is the working pressure head at the nozzle, and CD is the discharge coefficient, $= 0.98$.

6.2.4 Droplet Travel Distance

The simulation was performed by using Matrix Laboratory (MATLAB R2014a) software to predict the droplet size traveling distance, as shown in Fig. 6.1. The horizontal distance between the nozzle exit and the droplet landing point was simulated as the droplet travel distance. Droplet travel distance under no wind condition is undisturbed, and thus a characteristic of the droplet size for a given configuration. Droplet travel distance was simulated by substituting Eqs. 6.4–6.6 into Eqs. 6.1 and 6.2, for the droplet distribution. The input data were nozzle sizes, pressure, Trajectory distance, riser height, orifice coefficient, wind speed. A 0.1 mm droplet diameter increment was used starting from 0.1 mm to maximum droplet diameter. Finally, the model gives the droplet diameter and its distance from sprinkler as an output.

6.2.5 Estimation of the Droplet Size Distribution

Several sprinkler irrigation mathematical models considering droplet distribution have been developed in the last [12, 28, 25–31] presented a simulation scheme based on obtaining drop size distribution from the sprinkler radial water curve for a given sprinkler pressure combination under no wind conditions. In this study, [17] empirical model was adopted for the simulation. The reason for using the model was that it has been found to compare very well the inaccuracy to the well-known Upper Limit Log-Normal (ULLN) distribution model. The exponential model can be expressed as:

$$P_v = \left(1 - e^{-0.693\left(\frac{D}{D50}\right)^n}\right) \times 100 \quad (6.10)$$

where P_v is the percentage (%) of the total drops that are smaller than *D*; *D* is drop diameter (mm); D_{50} is volume mean drop diameter (mm); *n* is the dimensionless exponent.

$$D_{50} = a_d + b_d\, R \quad (6.11)$$

$$n = a_n + b_n\, R \quad (6.12)$$

The regression coefficients used for estimating the drop size distribution parameters for the dynamic fluidic sprinkler with a small round nozzle (5 mm) are as follows:

$$a_d = 0.29;\ b_d = 12{,}000;\ a_n = 2.04;\ b_n = -1400$$

where *R* is the ratio of the nozzle diameter to the pressure at the base of the sprinkler device.

6.2.6 Experimental Procedure

The sprinkler used in this study was specifically manufactured as an experimental sample by the Research Center of Fluid Machinery Engineering and Technology (Jiangsu University, Zhenjiang City, Jiangsu Province, China). The test nozzles were self-designed and locally machined using a wire-cut electric discharge machining process. The inlet diameter of the nozzle was set as 15 mm, while the outlet diameters were chosen as 4, 5, 6, and 7 mm and the shape of the nozzle was circular. Experiments were carried out in the sprinkler irrigation laboratory of Jiangsu University, China. The experiment setup is schematically presented in Fig. 6.2. The diameter of the circular shaped indoor laboratory was 44 m and height of 18 m. The materials used for the experiment include; centrifugal pump, electromagnetic flow meter, and piezometer, valve, and dynamic fluidic sprinkler. The sprinkler was mounted on a height of 1.9 m from the ground, with an elevation angle of 23°. The riser was at an angle of 90° to the horizontal from which 0.9 m above the ground. As

Fig. 6.2 Experimental setup in the indoor laboratory

shown in Fig. 6.2, water was drawn from the reservoir through the main pipe and ejected from the sprinkler. The experiment lasted for 4 h and the water temperature was 3 °C. Before the test was undertaken, the sprinkler was operated for several minutes to standardize the environmental conditions. A radial line was marked on the ground, extending from the sprinkler to the last observation point. Droplet sizes were determined using a 2D-Video Distrometer technique. It has the following specification, drop diameter measurement range from 0.125 to 6.5 mm with an increment of 0.125 mm and the measuring area is 1 m long, 1 mm wide with a thickness of 0.2 m and it was manufactured by Oanneum Research Digital-Institute of Information and Communication Technologies Steyrergasse 17, A-8010 Graz (Austria/Europe). The working principle is that two CCD line scan cameras face the opening of the lighting units. The object in the measurement area (determined by the cross-section of the two light paths viewed from above) blocks the light and is detected as shadows by the cameras. Further optical elements of the light paths, which have been omitted from this picture for the sake of simplicity, are two mirrors and a pair of slit plates that can contribute to the compact dimensions of the device and its insensitivity concerning spray. Each camera contains a small embedded computer that is responsible for handling the data capture process, the analysis of the data and its conversion and compression into a format suitable for further processing and transporting to the indoor user terminal. The droplet measurement was carried out at 2 m intervals along the radial direction of the sprinkler under a working pressure of 150, 200, 250 and 300 kPa. The sprinkler was allowed to spray over the measurement area a least five minutes to ensure a sufficient number of drops. A minimum of 100,000 drops were produced by the indoor user terminal only 92% of the drop sizes were analyzed after filtering.

6.2.7 Model Verification

To verify the model output, the predicted values were correlated to the measured values. A linear regression model of $Y = \beta + \beta_1 X$ was established with the predicted droplet diameter as the dependent variable (Y) and the observed droplet diameter as the independent variable (X). If the regression model is an ideal predictor of droplet diameter, the linear regression constants (β) and ($\beta 1$) will be equal to 0 and 1, respectively. Reference [30] pointed out that the values or R^2 (coefficient of determination) varies between 0 and 1 and provides an index of goodness of model fit. If the R^2 value is greater than or equal to 0.90, at least 90% of the variability is explained. This is generally considered to be very appropriate. On the other hand, the R^2 value of 0.80 is considered a good fit. An R^2 value as low as 0.60 is sometimes considered acceptable or even good. The evaluation of a linear model of different nozzles is based on values of β, β_1, R^2, R, and the standard error of estimation (Γ) which is defined as follows:

$$\Gamma = \sqrt{\frac{\sum_{i=n}^{i=n} \left(\Pi_m - \Pi_p\right)^2}{n}} \tag{6.13}$$

where, Π_m = measured droplet diameter, (mm); Π_p = predicted droplet diameter, (mm); Γ = standard error of estimation; n = number of observation.

The R^2 and Γ (standard error of estimate linear model) indicate the scatter points about the regression equation. R (correlation coefficient) indicates the degree of association between the observed and predicted values. To assist further in this evaluation, another index, known as the coefficient of efficient (ϖ) was used. This coefficient was proposed by [32], (1970) and used by [33]. If R and ϖ are close to each other, the model is free from any bias all or part of the data. (ϖ) is defined below as:

$$\varpi = \frac{\sum_{i=n}^{i=n} \left(\Psi_{oi} - \overline{\Psi_{oi}}\right)^2 - \sum_{i=n}^{i=n} \left(\Psi_{oi} - \Psi_p\right)}{\sum_{i=n}^{i-n} \left(\Psi_{oi} - \overline{\Psi_{0i}}\right)^2} \tag{6.14}$$

where ϖ = coefficient of efficient; n = number of observations; ψ_{oi} = value of observed measurements, (mm), ψ_p = value of predicted measurements, (mm), $-\overline{\Psi_{oi}}$ = average observed value, (mm).

6.3 Results and Discussion

Table 6.1 presents arithmetic mean droplet size diameter (mm) for different nozzle sizes and operating pressures along with the throw. Within a certain distance from the sprinkler, the average droplet diameter (arithmetic, volume, and median) usually increased with the increase of distance. Between 2 and 10 m, the volumetric drop diameter increased by 7.1%. Moreover, small water droplets were concentrated near the sprinkler, resulting in an average volumetric and median diameter less than 1 mm, while a volumetric and median diameter of more than 5.5 mm can be observed at a distance of 12 m. Similar findings were previously reported by many authors [34–36].

6.3.1 Comparison of the Measured Versus Predicted Droplet Size Diameter

Figure 6.3 shows a graphical comparison of droplet size diameters measured and predicted at 150 kPa for different nozzle sizes 4, 5, 6, and 7 mm. In general, the value of β_1 is close to 1 and β close to zero, accompanied by low λ and high R^2, R, and ϖ values, which indicates satisfactory prediction by the model. As the slope, β_1, and the

Table 6.1 Arithmetic mean droplet size diameter (mm) for different nozzle sizes and operating pressures

Pressure (kPa)	Nozzle Size (mm)	Distance from Sprinkler (m)					
		2	4	6	8	10	12
150	4	0.35	0.74	1.3	1.5	2.6	2.81
	5	0.78	0.94	1.36	2.48	3.03	3.11
	6	0.65	0.96	1.53	2.31	3.0	3.2
	7	0.51	0.83	1.5	2.4	2.45	2.7
200	4	0.31	0.75	1.23	1.31	2.36	2.489
	5	0.46	0.88	1.3	1.8	2.8	2.93
	6	0.45	0.8	1.32	1.75	2.37	2.51
	7	0.51	0.83	1.36	1.96	2.39	2.41
250	4	0.3	0.72	0.92	1.11	1.53	2.36
	5	0.42	0.76	1.2	1.72	2.64	2.67
	6	0.43	0.76	1.24	1.52	2.23	2.45
	7	0.5	0.8	1.35	1.68	2.31	2.35
300	4	0.26	0.68	0.85	0.98	2.03	2.33
	5	0.41	0.77	1.08	1.60	2.12	2.34
	6	0.37	0.78	1.04	1.25	2.07	2.31
	7	0.48	0.8	1.05	1.63	2.2	227

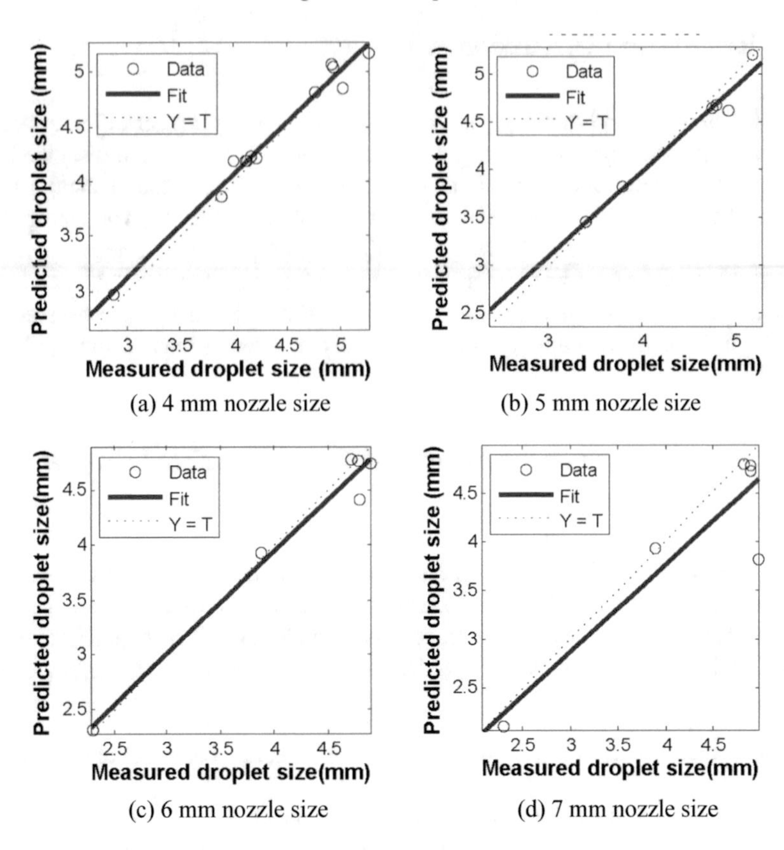

Fig. 6.3 A graphical comparison of the measured versus predicted droplet size diameter for different 4, 5 and 6 mm

intercept β are significantly different from 1.0 and 0, respectively, at the 99% level of confidence, a bias exists within the model estimation. This deviation oscillates between over and less estimation which depends mainly on β and β_1 values. Table 6.2 shows the evaluation results and statistical parameters of the droplet diameter.

Through a comprehensive evaluation of the four kinds of nozzles, it can be found that the R^2 value of all nozzle sizes is greater than 0.92, and the ϖ value is close to R^2. β and β_1 are close to 1 and 0, respectively. Furthermore, R^2 values are high, less difference between R^2 and ϖ, and Γ values are minimal. In general, the correlation between the observed and predicted droplet diameter values for all the nozzle sizes is satisfactory. This shows that the output of the model is suitable, and the deviation in the nozzle can be attributed to the experimental error, the change of the manufacturer, and uncalculated factors.

Table 6.2 Indices of the different orifice shapes in predicting droplet diameter

Parameter	Nozzle size (mm)			
	4	5	6	7
n	41	41	41	41
β	0.852	0.843	0.841	0.597
β_1	0.22	0.27	0.322	0.519
ϖ	0.931	0.981	0.967	0.943
R	0.964	0.984	0.9831	0.910
R^2	0.972	0.9892	0.986	0.956
Γ	0.214	0.192	0.23	0.215

6.3.2 Comparison of the Measured Versus Predicted Droplet Sizes for Different Pressures

Figure 6.4 presents a comparison between the measured droplet diameter and the predicted droplet diameter at different working pressures of 150, 200, 250, and 300 kPa. The results show that β1 is close to 1, β is close to 0, with low and high R2, R and ϖ values, the prediction results of the model are satisfactory. At the 99% confidence level, the slope β1 and intercept β are not significantly different from 1.0 and 0, respectively, so the model estimation is biased. This deviation oscillates between over and less estimation, which are mainly dependent on the values of β and β1. The evaluation results and statistical parameters of droplet diameter are given in the Table 6.3.

Through a comprehensive evaluation of the four pressure indexes, it can be found that R2 values for all sprinkler base pressures are greater than 0.91 and values are close to R2. β and β1 are close to 1 and 0, respectively. Besides, R2 values are high, less difference between R2 and, and values are minimal. In general, the correlation between the observed and predicted droplet diameters at 150 kPa, 200 kPa, and 250 kPa is more satisfactory than that of 300 kPa.

6.3.3 Comparison Between Other Simulated Travel Distance

Figure 6.5 shows the comparative analysis of the droplet travel distance between model and our model. It is clear from Fig. 6.5 that our model is consistent with model for the difference in the size range of large and small droplets. However, our model differs from [5]. This difference is mainly due to differences in the operating parameters used in the simulation.

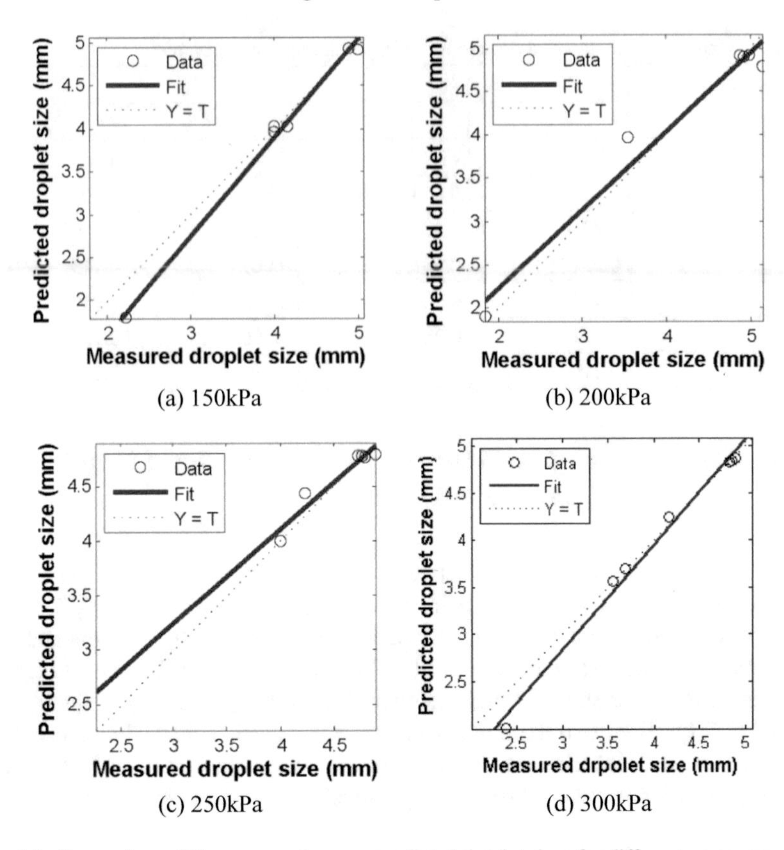

Fig. 6.4 Comparison of the measured versus predicted droplet sizes for different pressures

Table 6.3 Indices of the different sprinkler base pressures in predicting droplet diameter

Parameter	Pressure (kPa)			
	150	200	250	300
n	20	20	20	20
β	0.862	0.823	0.841	0.697
β_1	0.245	0.255	0.311	0.418
ϖ	0.971	0.967	0.965	1.246
R	0.964	0.953	0.937	0.912
R^2	0.981	0.975	0.973	0.954
Γ	0.114	0.182	0.20	0.218

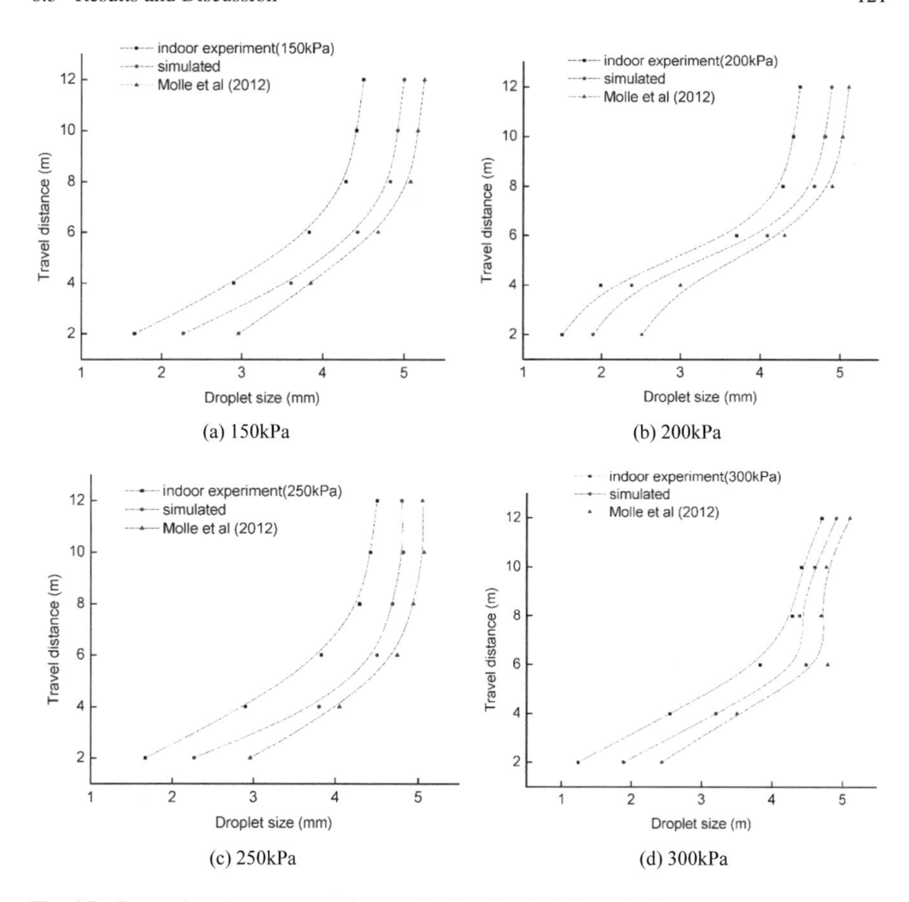

Fig. 6.5 Comparison between experiment, simulated and Molle et al. [5]

6.3.4 Compare the Droplet Size Distribution Model Prediction in Zero and Windy Conditions

The computer model was used to simulate droplet travel distances from the sprinkler for three wind speeds with downwind direction, and zero wind conditions are compared in Fig. 6.6. The droplet with a diameter of less than 1 mm traveled farthest. In the range of 1.5–5.5 mm, the traveled distance increases with the increase of droplet size and wind speed. The drift distance is the difference between the travel distance of the droplets in the same nozzle pressure configuration under the conditions of no wind and wind. Figure 6.6 is showing the extent of drift decreases as droplet size increases.

This highlights that the degree of drift is relatively more sensitive to the change of the size area of small droplets than that of large droplets, which makes the size of small droplets more prone to wind drift. Therefore, if the droplet distribution in a spray is seriously skewed to smaller droplet sizes, the distribution pattern can easily

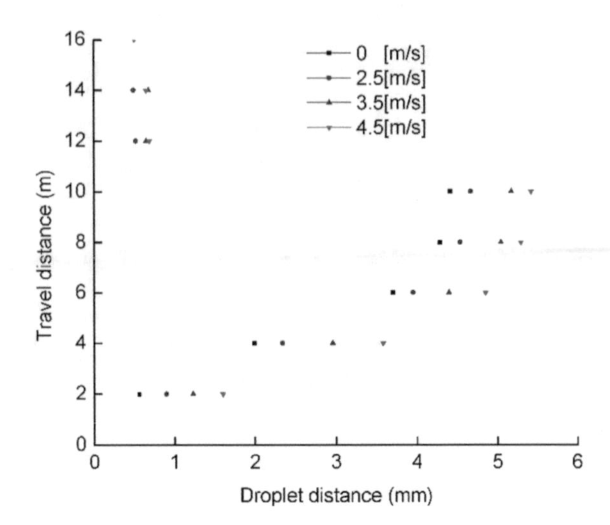

Fig. 6.6 Effect of wind speed on droplet size distribution compared with zero wind conditions at constant pressure (150 kPa)

be distorted under the influence of wind. The drift distance increases with the increase of wind speed. From Fig. 6.6, it is important to note that smaller diameter (0.5 to 1 mm) droplets were widely drifted compared to larger droplets (1.5–5.5 mm). For example, when the condition was 300 kPa and 2.5 m/s, the droplet between 0.5 mm and 3.94 mm did not exceed the characteristic wetting radius even though they were drifted. Only droplets with an average diameter of 4.45 mm, a frequency of 0.92%, as well as droplets with an average diameter range of less than 0.2 mm and a frequency of less than 3% of the total number of droplets, moved outside the wetting radius.

The remaining droplets have a higher probability of distorting the distribution pattern. This observation is particularly important because it partially answers the questions raised by [32, 37], distinguishing between water droplets that may cause loss of wind drift and high probability water droplets that only distort the distribution pattern. Although larger droplets account for only a small part of the number of droplets in all droplet distributions under consideration, they constitute a high percentage of loss if they are wind drifted due to their larger size per droplet. For example, at a constant working pressure of 300 kPa, at wind speeds of 3.5 m/s and 4.5 m/s, 20% and 32% of the total volume exceeded the wet radius, respectively. These are larger droplets (3.9–4.5 mm), accounting for 3.6% and 6.3% of the total number of distributed droplets, respectively. Therefore, the percentage of large droplets in the distribution spectrum is not only significant for predicting droplet effects [38, 39], but also more important for estimating wind drift losses, as they are likely to fall outside the wetting radius [40].

References

1. Wrachien D, Lnrenzini G (2006) Modeling jet flow and losses in sprinkler irrigation overview and perspective of a new approach. Biosys Eng 94:297–309
2. Silva WLC, Larry GJ (1988) Modeling evaporation and microclimate change in sprinkler irrigation I. Model formulation and calibration. Trans ASAE 31:1481–1486
3. Liu JP, Liu WZ, Bao Y, Zhang Q, Liu XF (2017) Drop size distribution experiments of gas-liquid two phase's fluidic sprinkler. J Drain Irrig Mach Eng 35:731–736
4. Zhu XY, Zhang AY, Zhang LG, Shi YJ, Jiang N (2021) Research on atomization performance of low-pressure atomization nozzle. J Drain Irrig Mach Eng 39:210–216
5. Molle B, Tomas S, Hendawi M (2012) Evaporation and wind drift losses during sprinkler irrigation influenced by droplet size distribution. Irrig Drain 61:240–250
6. Liu JP, Li T, Zhang Q (2021) Experimental study on influence of flow channel structure on hydraulic performance of low-pressure rotary sprinkler. J Drain Irrig Mach Eng 39:312–317
7. Zhu XY, Fordjour A, Yuan SQ, Dwomoh F, Ye DX (2018) Evaluation of hydraulic performance characteristics of a newly designed dynamic fluidic sprinkler. Water 10:1301
8. Zhu XY, Yuan SQ, Jiang JY, Liu JP, Liu XF (2015) Comparison of fluidic and impact sprinklers based on hydraulic performance. Irrig Sci 33:367–374
9. Shi YJ, Zhu XY, Hu G, Zhang AY, Li JP (2021) Effect of water distribution on different working conditions for sprinkler irrigation. J Drain Irrig Mach Eng 39:318–324
10. Hills DJ, Yuping G (1989) Sprinkler volume mean droplet diameter as a function of pressure. Trans ASAE 32:471–476
11. Li J (1997) Effect of pressure and nozzle shape on the characteristics of sprinkler droplet spectra. J Agric Eng Res 66:15–21
12. Solomon KH, Kincaid DC, Bezdek JC (1985) Drop size distributions for irrigation spray nozzles. Trans ASAE 28:1966–1974
13. Chen D, Wallender WW (1985) Droplet size distribution and water application with low-pressure sprinklers. Trans ASAE 28:511–516
14. Zhu X, Yuan S, Liu J (2012) Effect of sprinkler head geometrical parameters on the hydraulic performance of fluidic sprinkler. J Irrig Drain Eng 138:1943–4774
15. Zhang L, Zhou W, Li DX (2019) Research progress in irrigation mode selection of high-efficiency water-saving agriculture. J Drain Irrig Mach Eng 37:447–453
16. Yao JC, Wang XK, Zhang SC, Xu SR, Jin BB, Ding SW (2021) Orthogonal test of important parameters affecting hydraulic performance of negative pressure feedback jet sprinkler. J Drain Irrig Mach Eng 39:966–972
17. Xue ZL, Wang XK, Fan ED, Xu SR, Wang X, Zhang CX (2020) Structure optimization and test of three-way pulse jet tee. J Drain Irrig Mach Eng 38:751–756
18. Xu SR, Wang XK, Xiao SQ, Fan ED (2019) Numerical simulation and optimization of structural parameters of the jet-pulse tee. J Drain Irrig Mach Eng 37:270–276
19. Liu J, Zhu X, Yuan S, Fordjour A (2019) Modeling the application depth and water distribution of a linearly moving irrigation system. Water 10:1301. http://doi:11287/w11040827
20. Li YB, Liu JP (2020) Prospects for development of water saving irrigation equipment and technology in China. J Drain Irrig Mach Eng 38:738–742
21. Seginer I (1965) Tangential velocity of sprinkler drops. Trans ASAE 3, 90–93
22. Teske ME, Thistle HW, Londergan RJ (2011) Modification of droplet evaporation in the simulation of fine droplet motion using AGDISP. Trans ASABE 54:417–421
23. Xiang QJ, Xu ZD, Chen C (2018) Experiments on air and water suction capability of 30PY impact sprinkler. J Drain Irrig Mach Eng 36:82–87
24. Li J, Kawano H, Yu K (1994) Droplet size distributions from different shaped sprinkler nozzles. Trans ASAE 37:1871–1878
25. Liu JP, Liu XF, Zhu XY, Yuan SQ (2016) Droplet characterization of a complete fluidic sprinkler with different nozzle dimensions. Biosyst Eng 65:2–529
26. Song HB, Yoon SH, Lee DH (2000) Flow and heat transfer characteristic of a two dimensional oblique wall attaching offset jet. Int J Heat Mass Transf 43:2395–2404

27. Li H, Tang P, Chen C, Zhang ZY, Xia HM (2021) Research status and development trend of fertilization equipment used infertigation in China. J Drain Irrig Mach Eng 39:200–209
28. Fordjour A, Zhu X, Yuan S, Dwomoh F, Zakaria I (2020) Numerical simulation and experimental study on internal flow characteristics in the dynamic fluidic sprinkler. Appl Eng Agric 36:61–70
29. Fukui Y, Nakanishi K, Okamura J (1980) Computer evaluation of sprinkler irrigation uniformity. Irrig Sci 2:23–32
30. DeBor DW, Monnens MJ, Kincaid DC (2011) Measurement of sprinkler droplet size. Appl Eng Agric 17:11–115
31. Gregory JM, Fedler CB (1986) Model evaluation and research verification (MERV). ASAE Paper No. 1986, 86, 5032
32. Liu JP, Xu JE, Li T, Zaman M (2021) Relationship between solar energy and sprinkler hydraulic performance of solar sprinkler irrigation system. J Drain Irrig Mach Eng 39(6):637–642
33. Sharaf GA (2003) Evaluation of pressure distribution and lateral flow rates along drip tape lateral. J Agric Eng Res 20:542–556
34. Hu G, Zhu XY, Yuan SQ, Zhang LG, Li YF (2019) Comparison of ranges of fluidic sprinkler predicted with BP and RBF neural network models. J Drain Irrig Mach Eng 37:263–269
35. Seginer L, Kantz D, Nir D (1991) The distortion by the wind of the distribution patterns of single sprinklers. Agric Water Manag 19:314–359
36. Edling RJ (1985) Kinetic energy, evaporation and wind drift of droplets from low pressure irrigation nozzles. Trans ASAE 28:1543–1550
37. Nash JE, Sutcliffe JV (1970) River flow forecasting through conceptual models. I.A. Discuss. Princ J Hydrol 10:282–324
38. Forst KR, Schwalen HC (1960) Evapotranspiration sprinkler irrigation. Trans ASAE 3(18–20):24
39. Zhu X, Lewballah JK, Fordjour A, Jiang X, Liu J, Ofosu SA, Dwomoh FA (2021) Modelling of water drop movement and distribution in no wind and windy conditions for different nozzle sizes. Water 13(21):3006. https://doi.org/10.3390/w13213006
40. Li J, Hiroshi K (1995) Simulating water-droplet movement from noncircular sprinkler nozzles. J Irrig Drain Div ASCE 121:152–185

Chapter 7
Review of Intelligent Sprinkler Irrigation Technologies for Autonomous and Remote Sensing System

Abstract The aim of this chapter is to review the needs of soil moisture sensors in irrigation, sensor technology and their applications in irrigation scheduling and, discussing prospects. On board and field-distributed sensors can collect data necessary for real-time irrigation management decisions and transmit the information directly or through wireless networks to the main control panel or base computer. Communication systems such as cell phones, satellite radios, and internet-based systems are also available allowing the operator to query the main control panel or base computer from any location at any time. Selection of the communication system for remote access depends on local and regional topography and cost. Traditional irrigation systems may provide unnecessary irrigation to one part of a field while leading to a lack of irrigation in other parts. New sensor or remote sensing capabilities are required to collect real time data for crop growth status and other parameters pertaining to weather, crop and soil to support intelligent and efficient irrigation management systems for agricultural processes. Further development of wireless sensor applications in agriculture is also necessary for increasing efficiency, productivity and profitability of farming operations.

Keywords Precision agriculture · Data management · Autonomous system

7.1 Introduction

There has been a vast growth in the demand for water which has proved to be a cause of concern in irrigating agricultural fields. Agriculture is the major user of fresh water, consumes 70% of the fresh water, i.e. 1500 billion m^3 out of the 2500 billion m^3 of water is being used each year [1–4]. It is estimated that 40% of the fresh water that is used for agriculture in developing countries is lost either by evapotranspiration, spills or absorption by the deep layers of soil beyond the reach of roots. The problem of agricultural water management is today widely recognized as a major challenge that is often linked with development issues. Many freshwater resources have been degraded by agricultural activity, through over-exploitation, contamination with nutrients and salinization [5–8]. Different methods of irrigation

X. Zhu et al., *Dynamic Fluidic Sprinkler and Intelligent Sprinkler Irrigation Technologies*, Smart Agriculture 3, https://doi.org/10.1007/978-981-19-8319-1_7

are in use like drip irrigation, sprinkler irrigation, etc. to tackle with the water wastage problem in traditional methods like flood irrigation and furrow irrigation [9–13].

Productivity of agricultural fields varies for many reasons. The variability includes topographic relief, changes in soil texture, tillage and compaction, fertility differences, localized pest distributions and various irrigation system characteristics [14–18]. The effects of different sources of variability on management can be additive and interrelated. In this regard, recent advances in communications and microprocessors have led to the general implementation of site-specific water application systems by self-propelled and linear move sprinkler irrigation systems [19–22]. Designing of a suitable site-specific irrigation system could be complicated and challenging because it needs to address many causes of the variations existing in each field including the system capabilities that may be needed in achieving the desired management level, constraints inherent in the currently existing equipment and the general management philosophy of the owner/operator (decision support). These considerations are not mutually exclusive, but they do not lend themselves well to categorization. These issues are discussed in more detail by several researchers [23–27]. McCarthy et al. [28] developed a predictive-adaptive control model for site-specific irrigation water application of cotton using a center pivot. Various simulation models were used to evaluate alternative irrigation control options across a range of crop management and environmental conditions. The authors concluded that while the framework accommodated a range of system control strategies, further work is necessary to explore procedures for using data with a range of spatial and time scales.

Decision support systems should be holistic approaches to crop irrigation. Within the decision support program structure, the irrigator predefines the criteria and guidelines to be used by the software structure and simulation models in making basic decisions to be implemented by a microprocessor-based control system. Results of geo-referenced grid sampling of soils, yield maps and other precision agriculture tools can also be major components in defining rules for these management systems. These 'rules' are used as the basis for analysing and interpreting the data from real time data networks, remote sensing, irrigation monitoring systems, agronomic and other information used to provide direction and implement of basic commands [29, 30]. Decision support systems can also include instructions for chemigation (e.g., nitrogen fertilizer) and provide alerts (e.g., insects, diseases) to the grower based on output from established models using real-time environmental data. In short, decision support provides more management flexibility by implementing short term, routine commands to direct irrigation schedules and other basic operations, which frees the irrigator to concentrate on managing other areas to minimize risk and reduce costs [31, 32].

Vast simulation models or integrated approaches were used to evaluate alternative irrigation management options across a range of crop and environmental conditions [33–35]. These integrated approaches require the integration of various sensor systems (on the irrigation machine and in the field), hardware, controllers and computing power. The maximum benefits will be derived from a decision support system when the plant condition in selected areas of a field is monitored by some

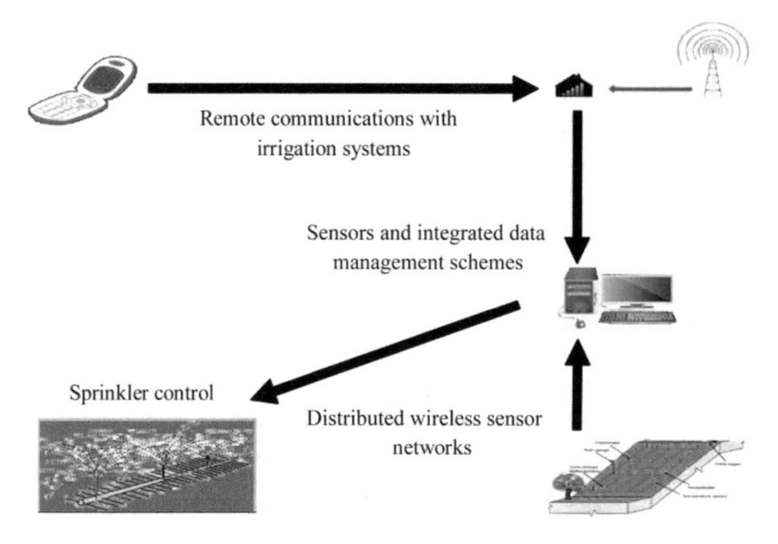

Fig. 7.1 Relationship of autonomous sensor irrigation management technologies

means to improve overall system management. Monitoring systems can be field-based measurements or remotely sensed or an integrated mix of several sensor systems [36].

In this research, the techniques used to gain information about autonomous sensor irrigation management technologies will be explained. This includes information collecting using wireless sensor networks, remote sensor connection, data management schemes and target controlling. There are rare reviews on the topic of autonomous sensor irrigation management in the existing literature and this study aims to fill the gap and provide a more up to date and thorough review, featuring many new developments in irrigation management and an extension to the application of latest sensor technology in agricultural engineering. The relationship of the sections of this review paper is as illustrated in Fig. 7.1.

7.2 Autonomous Sensor Irrigation Management Technologies

7.2.1 Remote Access and Communications

This section mainly focuses on the components used for remotely controlling or monitoring sprinkler irrigation systems through computers and accessories. Many of these methods are being marketed by the manufacturers of this equipment and include cell phones, RF radios, and satellite radio communications for relatively basic monitoring and control of the systems. Hybrid systems relying on internet to connect computers at or near the site are combined with wireless RF systems for the link to the machine [37, 38].

Each manufacturer has developed unique hardware and software that allow the owner to access the main control panel to determine system status including travel direction, application depth, and field position. More sophisticated software provided by an office base station uses visualization software to allow the owner to see year-to-date summaries of water and chemical application events [39–41]. Shock et al. [42] used radio transmission for soil moisture data from data loggers to a central computer logging site.

Researchers within the Soil and Crop Science and Civil Engineering Department at Colorado State University have created WISE for agricultural producers, irrigation managers, and research scientists [43, 44]. The tool resides on a cloud based platform of the environmental Risk Assessment and Management System, and uses the soil water balance (SWB) approach to assist users by providing recommended irrigation amounts for individual fields. Mandatory setup of a field can only be completed on a web browser via a computer. The difficulty of outlining field shapes on a smart phone or tablet prohibits users from using the full version of WISE on those devices; therefore, the mobile version does not possess full capabilities of WISE. Instead, users can easily view their soil moisture profile, add irrigations, precipitation, and observed soil deficit values once the project and field are set up using a web browser on a personal computer [45–47].

In GSM based automated irrigation control using a gun sprinkler [48, 49] mentioned about using automatic microcontroller based rain gun irrigation system in which the irrigation will take place only when there will be intense requirement of water that save a large quantity of area. Mobile phones have almost become an integral part of us serving multiple needs of humans. Mobile phone applications make use of the GPRS feature of mobile phone as a solution for irrigation control system. These systems cover lower range of agriculture land and not economically affordable. The system uses GSM to send message and an android app is used and it can notify the farmer to overcome under irrigation, over irrigation that causes leaching and loss of nutrient content of soil [50, 51].

In GSM based automatic irrigation control system for efficient use of resources and crop planning by using an Android Mobile phone [52, 53] states feature of their system supports water management decision, used for monitoring the whole system with GSM(RS-232) module. The system continuously monitors the water level (water level sensor) in the tank and provide accurate amount of water required to the plant or tree (crop). The system checks the temperature, and humidity of soil to retain the nutrient composition of the soil managed for proper growth of plant. This system is low cost and effective with less power consumption using sensors for remote monitoring and controlling devices which are controlled via SMS using a GSM using android mobile.

In irrigation control system using Android mobile apps and GSM for efficient use of water and power, autonomous irrigation system uses valves to turn the pump ON and/or OFF. These valves may be easily automated by using controllers. Automating farm or nursery irrigation allows farmers to apply the right amount of water at the right time, regardless of the availability of labor to turn valves on and off. In addition, farmers using automation equipment can reduce runoff from over watering saturated soils, avoid irrigating at the wrong time of day, which will improve crop performance by ensuring adequate water and nutrients when needed. Those valves may be easily automated by using controllers [54, 55].

Selection of the communications system for remote access depends on topography and cost relative to other methods. Cell phone systems with modems at the control panels are the least costly and probably the most common. Satellite radio communications are often preferable when there are large topographic differences that limit cell phone service [56, 57]. Higher powered, licensed, radio systems (e.g., 5–10 W) with data modems may also be an option but may also be affected by topographic relief. Repeater stations for radio frequency systems can also be quite expensive, especially if there is a need to communicate long distances over diverse topography. These additions to the existing on- board control capabilities of center pivot panels make site-specific irrigation a reality for irrigation zones less than the 100 m^2. The main considerations remaining include the development of decision support systems that maximize the value of the applied water or chemical based on field-specific information and the cost recovery potential of the cropping system since system costs up to $20 000 are possible when there is many management zones along the system length.

7.2.2 Distributed Wireless Sensor Networks

In-field sensor-based irrigation systems offer the potential to support site-specific irrigation management that allows producers to maximize their productivity while saving water. The seamless integration of sensors, data interface, software design, and communications for site specific irrigation control using wireless sensor-based irrigation systems can also be challenging [58]. Electrical power needs are often a major consideration and solar panels are often used. A number of researchers have addressed the issues of interfacing sensors and irrigation control using several different approaches. Shock et al. [42] used radio transmission for soil moisture data from data loggers to a central data logging site where decisions were made and manually changed. Wall and King [59] explored various designs for smart soil moisture sensors and sprinkler valve controllers for implementing 'plug-and-play' technology, and proposed architectures for distributed sensor networks for site specific irrigation automation. They concluded that the coordination of control and instrumentation data is most effectively managed using data networks and low-cost microcontrollers. However, it is often not feasible to have in-field sensing stations that use wires to connect to a base station because of the cost, labour and maintenance, especially if the distances are greater than 10 m. Wires can also be damaged by farm equipment and small animals; and wires create more opportunity for lightning damage. In this regard, wireless data communication systems avoid many of these problems and provide dynamic mobility and easy relocation and replacement of stations. Radio frequency technology has been widely adopted in consumer's wireless communication products and provided opportunities to deploy wireless signal communication in agricultural systems. Adopting a standard interface for sensors and actuators allows reuse of common hardware and communication protocols such as communication interface and control algorithm software. Instrumentation and control standards for RS232 serial (voltage based) and RS485 (current based) communication protocols have been widely applied and well documented for integrating sensors and actuators, particularly in industrial applications. Two wireless protocols that are commonly used for this purpose are Bluetooth (802.15.1) (IEEE Std. 802.15.1, 2005) and ZigBee (802.15.4) (IEEE Std. 802.15.4a, 2007). Bluetooth and ZigBee (IEEE 802.11 standards) are designed for radio-frequency (RF) applications for mobile applications that require a relatively low data rate, long battery life, and good network security [60–63].

These are 'line-of-sight' (LOS) systems and crop canopies, small trees, and fences can interfere with transmissions. ZigBee is a low-cost, non-proprietary wireless mesh networking standard, which allows longer life with smaller batteries, and the direct-sequence spread spectrum (DS/SS) mesh networking provides high reliability. Bluetooth is a faster but more expensive standard than ZigBee, and uses spread spectrum modulation technology called frequency hopping (FH/SS) to avoid interference and ensure data integrity. ZigBee has lower power needs than Bluetooth, but it also transmits effectively over less distance (e.g., 30 m). Enhanced Bluetooth transmitters are available that can transmit up to 1 km. Bluetooth wireless technology has been adapted in sensing and control of agricultural systems [64, 65]. Zhang [66] evaluated Bluetooth radio in different agricultural environments, power consumption levels, and data transmission rates. He observed 1.4 m as an optimal radio height for maximum 44 m radio range and reported limitations of significant signal loss after 8 h continuous battery operation and 2–3 s of transmission latency with the increase of communication range. Oksanen et al. [67] used a PDA with Bluetooth to connect a GPS receiver for their open, generic and configurable automation platform for agricultural machinery. Lee et al. [68] explored an application of Bluetooth wireless data transportation of moisture concentration of harvested silage and reported a limitation of 10 m short range. However, the limitations reported by reviewed publications about Bluetooth applications in agricultural systems can be solved or minimized by system design optimization. The power shortage can be solved by using solar power that recharges the battery. The radio range and transmission latency can also be extensively improved by using an upgraded power class and antenna. The same techniques can be applied to Zigbee-based systems.

Drawbacks in using wireless sensors and wireless sensor networks include provision for ample bandwidth, existing inefficiencies in routing protocols, electromagnetic interference, interference by vegetation, radio range, sensor battery life, and synchronous data collection [69]. An immediate limiting factor in self-powered WSN operations is battery life, which can be addressed to some degree by decreasing the duty cycle of the sensor nodes. Researchers are also concentrating on RF communication protocols to increase the energy efficiency of a WSN by investigating algorithms for multi-path routing, data throughput and energy consumption, and by reducing

Table 7.1 Advantages and disadvantages of distributed Wireless Sensor Networks (WSN)

Advantages	Disadvantages	
Allows producers to maximize their productivity while saving water	Power needs are often a major consideration	Interference by vegetation
Provide dynamic mobility and easy relocation and replacement of stations	Provision for ample bandwidth	Synchronous data collection
Coordination of control and instrumentation data is most effectively	Existing inefficiencies in routing protocols	Sensor battery life Radio range
Managed using data networks and low-cost microcontrollers	Electromagnetic interference	Interference with radio propagation to crop canopy height

idle listening and collisions that occur during the medium access to realize power conservation [70, 71].

However, reducing quiescent current draw is typically a significant method for impacting battery longevity [72]. Other identified challenges specific to WSNs and agriculture include interference with radio propagation due to crop canopy height [73]. Andrade-Sanchez et al. [74] determined that power consumption and power output varied significantly among transceivers, and the average measure of signal strength as a function of distance resembled the shape of the theoretical prediction of path loss in free space. In addition, the received signal strength indication (RSSI) was influenced by the spatial arrangement of the network in both the vertical and horizontal planes in tests with line-of-sight. Signal obstruction issues relating to crop height and in-field equipment are inherently reduced when the moving sprinkler is used as the sensor platform; but infield sensors require manual adjustment above crop canopy.

The Table 7.1 summarizes the advantages and disadvantages of distributed Wireless Sensor Networks (WSN).

7.2.3 Sensors and Integrated Data Management Schemes

In this section, to determine the soil moisture content either in volumetric and gravimetric forms, various techniques can be employed, which can be categorized. These are classical and modern techniques for both the laboratory and in situ measurements. The classical soil moisture measurement techniques include thermo-gravimetric, calcium carbide neutron scattering, gypsum block and tensiometer methods [75–77]. While the modern techniques utilize soil resistivity sensor, tensiometer, infrared moisture balance and dielectric techniques like Time Domain Reflectometry (TDR), Frequency Domain Reflectometry (FDR) capacitance technique, heat flux soil moisture sensors, micro-electro mechanical systems and optical techniques [78, 79]. Estimation of water content based on sensor measurements provides real time, in situ measurements at a relatively affordable cost. Soil moisture sensors potentially provide the means to irrigate in accordance with the unique characteristics of a given crop in a given field. These sensors can be used as a 'stand-alone' method, or their use can be combined with the FAO method, or they can be used to complement irrigation management based on experience [80].

Automating farm or nursery irrigation systems allows farmers to apply the right amount of water at the right time, regardless of the availability of labor to turn valves on and off. It only allows the user to monitor and maintain the moisture level remotely irrespective of time. If the plants get water at the proper time then it helps to increase the production from 25 to 30% [81, 82]. The main aim of this paper is to provide automatic irrigation to the plants which helps in saving money and water. Irrigation by help of freshwater resources in agricultural areas has a crucial importance. Traditional instrumentation based on discrete and wired solutions, presents many difficulties on measuring and control systems especially over the large geographical areas. If different kinds of sensors (that is, temperature, humidity, etc.) are involved in such irrigation in future works, it can be said that an internet based remote control of irrigation automation will be possible. The developed system can also transfer fertilizer and the other agricultural chemicals (calcium, sodium, ammonium, zinc) to the field with adding new sensors and valves. Cost effective solar power can be the answer for all our energy needs. Conserves electricity by reducing the usage of grid power and conserves water by reducing water losses [83, 84].

A conceptual system layout of distributed in-field Wireless Sensor Network (WSN) is illustrated in Fig. 7.2 [85]. Farmers can get the real-time information of their farmland by android app or through automatic SMS facility, for better crop management practices. Using this information, the farmers could be advised that when and how much to irrigate.

The system comprises of several components called 'nodes'. These are smart devices that are used to collect the application oriented data requirements. A sensor network performs three basic functions that is Sensing, Communication and Computation by using hardware, software and algorithm [86]. The nodes perform several roles and the distributed nodes that collect the information are called source node while the node that gathers the information from all source node is called the sink node or the gateway node. The sink node has high computing power. A source node also works as a routing node due to the requirement of multi hop routing. External memory is an optional module that could be needed in case of data storage requirement for local decision making. The in-field sensors monitor the field conditions of soil moisture, soil temperature, and air temperature. All in-field sensory data are wirelessly transmitted to the base station. The base station processes the in-field sensory data through a user-friendly decision-making program and sends control commands to the irrigation control station [87].

Dias et al. [88] developed a new single probe heat pulse sensor (SPHP), which was comprised of only one element, a n-p–n junction bipolar transistor, worked as both heating and temperature sensing elements. Xiao et al. [89] developed a

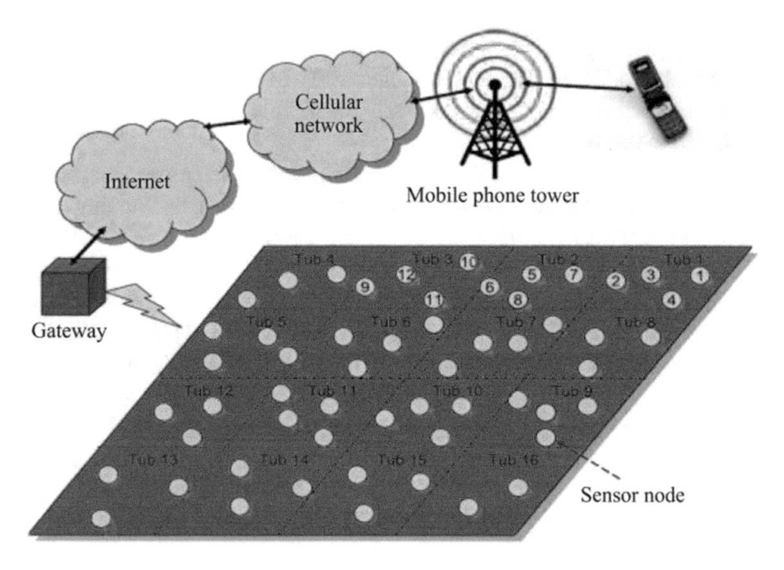

Fig. 7.2 Application of sensor network in field

wireless, integrated, frequency domain soil moisture sensor for paddy field (WFDSS) applications in china. This soil sensor was able to measure soil moisture content and water depth at the same time and transmitted the collected data wirelessly to a remote data management center.

7.2.4 Sprinkler Control Options

Center pivot systems, for example, operate on varying topography and often have a range in soil textures present under a single machine. Each of these factors represents a reason for using some sort of monitor/controller to manage water applications based upon need. Precision application, variable rate irrigation and site-specific irrigation are terms developed to describe water application devices with the goal of maximizing the economic and/or environmental value of the water applied via a moving irrigation system [90–92]. The most basic method to alter the water depth applied with a center pivot is to adjust the center pivot speed of travel based upon field soils or more frequently based upon field topographic features or different crops. Early developments provided a very limited set of controls to turn end guns on and off based upon field position. Other features included edge of field stops and stop-in-slot controls to cease irrigation due to obstructions or the completion of a complete rotation [93–96].

Programmable control panels allow adjusting the speed of travel multiple times during an irrigation event. This is accomplished by entering the field position in a 360° circle where the speed will be changed to apply more, less or no irrigation water. This approach could be used where portions of the field were planted to a different crop, but it lacked the flexibility necessary to supply water at rates required to meet management objectives of relatively small field areas with irregular shaped boundaries [97–99].

The individual sprinkler control of water application depth can be accomplished by using a series of on–off time cycles or as it has become known as 'pulsing' the sprinkler through on–off cycles [100, 101]. Reducing the on-time is effective at reducing both the application depth and the water application rate. Later efforts in Washington State involved equipping a center pivot with a custom built electronic controller to activate water operated solenoid valves in groups of 2–4 nozzles [102–104]. Normally open solenoids allowed system control with the assurance that irrigation water was applied even if the control system failed. Chávez et al. [105] reported that a remote irrigation monitoring and control system installed on two different linear move irrigation systems performed well. The systems proved to be highly flexible and capable of precision irrigation using a series of in-field and on-board wireless monitoring spread spectrum radios/sensors networks. Individual nozzle/solenoid valves were pulsed according to prescription maps. Deviations related to positioning of nozzles when irrigating were on average (2.5 ± 1.5) m due mainly to inherent DGPS inaccuracy. A variable flow sprinkler was developed for controlling irrigation water application by King and Kincaid and Liu et al. The variable flow sprinkler uses a mechanically-activated pin to alter the nozzle orifice area which adjusted the sprinkler flow rate over the range of 35% to 100% of its rated flow rate based upon operating pressure. The pin was controlled using either electric or hydraulic actuators. The main issue is that the wetted pattern and water droplet size distribution of the sprinkler changed with flow rate which created water application uniformity issues due to a change in sprinkler pattern overlap. Controlling irrigation water application depth can also be accomplished using multiple manifolds with different sized sprinkler nozzles to vary water and nitrogen application. These systems included 2–3 manifolds where simultaneous activation of one or manifolds served to adjust the water application rate and depth across a range of depths that is not possible with a single sprinkler package. Control of each manifold was accomplished using solenoid valves like those described for the pulsing sprinkler option above. As with any new technology, there are positives and negatives associated with each of these methods of controlling sprinkler flow rates. Certainly, long term maintenance by producers is an issue. However, the biggest factor limiting their use is their installation cost that ranges from around \$2000 for a system monitor to over \$20 000 for control of individual sprinklers.

7.3 Conclusions and Future Work

Electronic sensors, equipment controls, and communication protocols have been developed to meet the growing interest in site-specific irrigation systems. On-board and field-distributed sensors can collect data necessary for real-time irrigation management decisions and transmit the information through wireless networks to the main control panel or base computer. Equipment controls necessary to alter water application depth to meet the management criteria for relatively small management zones are now commercially available from irrigation system manufacturers and after-market suppliers. But decision systems for automatic control are incomplete. Selection of the communications system for remote access depends on local and regional topography and cost relative to other methods. Communication systems such as cell phones, satellite radios, and internet based systems allow the operator to query the main control panel or base computer from any location at any time. Recent developments in the center pivot industry have led to contractual relationships between after-market suppliers and irrigation system manufacturers that should support further development of site-specific application of water, nutrients and pesticides in the future.

However, the limitations reported in reviewed publications about Bluetooth applications in agricultural systems can be solved or alleviated by system design optimization. The power shortage can be solved by using solar power that recharges the battery. The radio range and transmission latency can also be extensively improved by using an upgraded power class and antenna. The same techniques can be applied to Zigbee-based systems. Considering a real need to improve the efficiency of irrigation systems and prevent the non-optimal use of water, the most promising and challenging research direction is to develop intelligent irrigation management systems which will enable farmers and other end users to optimize the use of water; for example, only irrigate where and when it is needed. This would provide site specific and time specific irrigation management to maximize the efficiency of the irrigation so reduce the use of water. This is a very complicated and challenging research topic as crop water productivity is affected not only by the environment and soil but also by what kind of crop and at what a stage. Whenever there is a change in local environment such as temperature and humidity, sensors are required to sense or observe these changes and the irrigation decision making system need to consider these changes and update irrigation strategies accordingly. New sensor or remote sensing capabilities are required to collect real time data for crop growth status and other parameters pertaining to weather, crop and soil to support intelligent and efficient irrigation management systems for agricultural processes. Further development of wireless sensor applications in agriculture is also necessary for increasing efficiency, productivity and profitability of farming operations.

References

1. Luo XW, Liao J, Hu L, Zhang Y, Zhou ZY (2016) Improving agricultural mechanization level to promote agricultural sustainable development. Trans CSAE 32(1):1–11 (in Chinese)
2. Giagnocavo C, Bienvenido F, Li M, Zhao YR, Sanchez-Molina JA, Yang XT (2017) Agricultural cooperatives and the role of organisational models in new intelligent traceability systems and big data analysis. Int J Agric Biol Eng 10(5):115–125

3. Wolfert S, Ge L, Verdouw C, Bogaardt MJ (2014) Big data in smart farming: a review. Agric Syst 153:69–80
4. Shah NG, Das I (2012) Precision irrigation sensor network based irrigation, problems, perspectives and challenges of agricultural water management. IIT Bombay, India, pp 217–232
5. Yuan SQ, Darko RO, Zhu XY, Liu JP, Tian K (2017) Optimization of movable irrigation system and performance assessment of distribution uniformity under varying conditions. Int J Agric Biol Eng 10(1):72–79
6. Thompson RB, Gallardo M, Valdez LC, Fernández MD (2007) Using plant water status to define threshold values for irrigation management of vegetable crops using soil moisture sensors. Agric Water Manag 88:147–158
7. Hsiao TC, Steduto P, Fereres E (2007) A systematic and quantitative approach to improve water use efficiency in agriculture. Irrig Sci 25:209–231
8. Sadeghi SH, Peters T, Shafii B, Amini MZ, Stöckle C (2017) Continuous variation of wind drift and evaporation losses under a linear move irrigation system. Agric Water Manag 182(3):39–54
9. Zhu M, Zhou XQ, Zhai ZF (2016) Research progresses in technological innovation and integration of agricultural engineering. Int J Agric Biol Eng 9(6):1–9
10. Keller J, Bliesner RD (2000) Sprinkler and trickle irrigation. The Blackburn Press, Caldwell
11. Peters RT, Evett SR (2007) Spatial and temporal analysis of crop stress using multiple canopy temperature maps created with an array of center-pivot-mounted infrared thermometers. Trans ASABE 50(3):919–927
12. Zhu XY, Peters T, Neibling H (2016) Hydraulic performance assessment of LESA at low pressure. Irrig Drain 65(4):530–536
13. Liu JP, Yuan SQ, Li H, Zhu XY (2016) Experimental and combined calculation of variable fluidic sprinkler in agriculture irrigation. Agric Mech Asia Africa Latin Am 47(1):82–88
14. Shankar V, Ojha CSP, Prasad KSH (2012) Irrigation scheduling for maize and Indian-mustard based on daily crop water requirement in a semi-arid region. Int J Civil Environ Eng 6:476–485
15. Lan YB, Thomson SJ, Huang YB, Hoffmann WC, Zhang HH (2010) Current status and future directions of precision aerial application for site-specific crop management in the USA. Comput Electron Agric 74(1):34–38
16. Li LH, Zhang XY, Qiao XD, Liu GM (2016) Analysis of the decrease of center pivot sprinkling system uniformity and its impact on maize yield. Int J Agric Biol Eng 9(4):108–119
17. Duan FY, Liu JR, Fan YS, Chen Z, Han QB, Cao H (2017) Influential factor analysis of spraying effect of light hose-fed traveling sprinkling system. J Drain Irrig Mach Eng 35(6):541–546 (in Chinese)
18. Lin YY, Zhang ZX, Xu D, Nie TZ (2016) Effect of water and fertilizer coupling optimization test on water use efficiency of rice in black soil regions. J Drain Irrig Mach Eng 34(2):151–156
19. Yan HJ (2005) Study on water distribution irrigation uniformity of center pivot and later move irrigation system based on variable rate technology. PhD Thesis, China Agricultural University, 2005, pp 95–96 (in Chinese)
20. Peters TR, Evett SR (2008) Automation of a center pivot using the temperature-time threshold method of irrigation scheduling. J Irrig Drain Eng ASCE 134(1):286–291
21. Sadler EJ, Evans RG, Stone KC, Camp CR (2005) Opportunities for conservation with precision irrigation. J Soil Water Conserv 60(6):371–379
22. Zhu XY, Yuan SQ, Liu JP (2012) Effect of sprinkler head geometrical parameters on hydraulic performance of fluidic sprinkler. J Irrig Drain Eng ASCE 138(11):1019–1026
23. Buchleiter GW, Camp C, Evans RG, King BA (2000) Technologies for variable water application with sprinklers In: Phoenix AZ. Evans RG, Benham BL, Trooien TP (eds) Proceedings of 4th decennial national irrigation symposium . November 14–16, 2000. ASAE, St. Joseph, MI. (Publication 701P0004), 2000; pp 316–321
24. Evans RG, Buchleiter GW, Sadler EJ, King BA, Harting GB (2000) Controls for precision irrigation with self-propelled systems. In: Phoenix AZ, Evans RG, Benham BL, Trooien TP (eds) Proceedings of 4th Decennial national irrigation symposium. November 14–16, 2000. ASAE, St. Joseph, MI. (Publication 701P0004), pp 322–331

25. Sadler EJ, Evans RG, Buchleiter GW, King BA, Camp CR (2000) Site-specific irrigation - management and control. In: Phoenix AZ, Evans RG, Benham BL, Trooien TP (eds) Proceedings of 4th decennial national irrigation symposium. November 14–16, 2000. ASAE, St. Joseph, MI. (Publication 701P0004), pp 304–315
26. Perry CD, Dukes MD, Harrison KA (2004) Effects of variable-rate sprinkler cycling on irrigation uniformity. ASABE Paper No. 041117. St. Joseph, MI: ASABE
27. Christiansen JE (1942) Irrigation by sprinkling. California agricultural experiment station. Bulletin 670. University of California, Berkeley, CA
28. McCarthy AC, Hancock NH, Raine SR (2010) VARIwise: a general-purpose adaptive control simulation framework for spatially and temporally varied irrigation and sub-field scale. Comput Electron Agric 70:117–128
29. Gowda PH, Chávez JL, Colaizzi PD, Evett SR, Howell TA, Tolk JA (2008) ET mapping for agricultural water management: present status and challenges. Irrig Sci 26:223–237
30. Mahan JR, Conaty W, Neilsen J, Payton P, Cox SB (2010) Field performance in agricultural settings of a wireless temperature monitoring system based on a low-cost infrared sensor. Comput Electron Agric 71:176–181
31. Gil E, Arnó J, Llorens J, Sanz R, Llop J, Rosell JR et al (2014) Advanced technologies for the improvement of spray application techniques in Spanish viticulture: an overview. Sensors 14(1):691–708
32. Darko RO, Yuan SQ, Liu JP, Yan HF, Zhu XY (2017) Overview of advances in improving uniformity and water use efficiency of sprinkler irrigation. Int J Agric Biol Eng 10(2):1–15
33. Peters RT, Evett SR (2005) Using low-cost GPS receivers for determining field position of mechanical irrigation systems. Appl Eng Agric 21(5):841–845
34. Liu JP, Yuan SQ, Li H, Zhu XY (2013) A theoretical and experimental study of the variable-rate complete fluidic sprinkler. Appl Eng Agric 29(1):17–24
35. Liu JP, Yuan SQ, Li H, Zhu XY (2013) Numerical simulation and experimental study on a new type variable-rate fluidic sprinkler. J Agric Sci Technol 15(3):569–581
36. Lan YB, Chen SD, Fritz BK (2017) Current status and future trends of precision agricultural aviation technologies. Int J Agric Biol Eng 10(3):1–17
37. Xuan CZ, Wu P, Zhang LN, Ma YH, Liu YQ (2017) Compressive sensing in wireless sensor network for poultry acoustic monitoring. Int J Agric Biol Eng 10(2):94–102
38. Camilli A, Cugnasca CE, Saraiva AM, Hirakawa AR, Correa PLP (2007) From wireless sensors to field mapping: anatomy of an application for precision agriculture. Comput Electron Agric 58(1):25–36
39. Wang P, Luo XW, Zhou ZY, Zang Y, Hu L (2014) Key technology for remote sensing information acquisition based on micro UAV. Trans CSAE 30(18):1–12 (in Chinese)
40. Ma HQ, Huang WJ, Jing YS (2016) Wheat powdery mildew forecasting in filling stage based on remote sensing and meteorological data. Trans CSAE 32(9):165–172 (in Chinese)
41. Zhu HP, Masoud S, Robert DF (2011) A portable scanning system for evaluation of spray deposit distribution. Comput Electron Agric 76:38–43
42. Shock CC, David RJ, Shock CA, Kimberling CA (1999) Innovative, automatic, low cost reading of Watermark soil moisture sensors. In: Proceedings of 1999 irrigation association technical conference, the irrigation association, falls Church, VA, 1999; pp147–152
43. Andales AA, Bauder TA, Arabi M (2014) A mobile irrigation water management system using a collaborative GIS and weather station networks. In: Ahuja LR, Ma L, Lascano R (eds) Practical applications of agricultural system models to optimize the use of limited water, advances in agricultural systems modeling. ASA, CSSA, and SSSA, Madison, WI, USA, 2014; pp 53–84
44. Zhang HH, Lan YB, Charles PCS, Westbrook J, Hoffmann WC, Yang CH (2013) Fusion of remotely sensed data from airborne and ground-based sensors to enhance detection of cotton plants. Comput Electron Agric 93:55–59
45. Willers JL, Jenkins JN, Ladner WL, Gerard PD, Boykin DL, Hood KB (2005) Site-specific approaches to cotton insect control. Sampling and remote sensing analysis techniques. Precis Agric 6:431–452

46. Song Y, Sun H, Li M, Zhang Q (2015) Technology application of smart spray in agriculture: a review. Intell Autom Soft Comput 21(3):319–333
47. Andrade-Sánchez P, Upadhyaya SK, Jenkins BM (2007) Development, construction, and field evaluation of a soil compaction profile sensor. Trans ASABE 50(3):719–725
48. Han XZ, Kim HJ, Jeon CW, Moon HC, Kim JH (2017) Development of a low-cost GPS/INS integrated system for tractor automatic navigation. Int J Agric Biol Eng 10(2):123–131
49. Kang F, Pierce FJ, Walsh DB, Zhang Q, Wang S (2011) An automated trailer sprayer system for targeted control of cutworm in vineyards. Trans ASABE 54(4):1511–1519
50. Mulla DJ (2013) Twenty five years of remote sensing in precision agriculture: Key advances and remaining knowledge gaps. Biosyst Eng 114(4):358–371
51. Piekarczyk J (2014) Application of remote sensing in agriculture. Precis Agric 13(1):69–75
52. Montoya FG, Gómez J, Cama A, Sierra AZ, Martínez F, de la Cruz JL et al (2013) A monitoring system for intensive agriculture based on mesh networks and the android system. Comput Electron Agric 99:14–20
53. Wu C, Tang Y, Tang LD, Chen J, Li K (2017) Characteristic parameter wireless monitoring system of hydraulic turbine based on android. J Drain Irrig Mach Eng 35(4):362–368 (in Chinese)
54. Huang YB, Thomson SJ, Brand HJ, Reddy KN (2016) Development of low-altitude remote sensing systems for crop production management. Int J Agric Biol Eng 9(4):1–11
55. Yu FH, Xu TY, Du W, Ma H, Zhang GS, Chen CL (2017) Radiative transfer models (RTMs) for field phenotyping inversion of rice based on UAV hyperspectral remote sensing. Int J Agric Biol Eng 10(4):150–157
56. Wang P, Zhang JX, Lan YB, Zhou ZY, Luo XW (2014) Radiometric calibration of low altitude multispectral remote sensing images. Trans CSAE 30(19):199–206 (in Chinese)
57. Liang Q, Yuan D, Wang Y, Chen HH (2007) A cross-layer transmission scheduling scheme for wireless sensor networks. Comput Commun 30:2987–2994
58. King BA, Wall RW, Wall LR (2000) Supervisory control and data acquisition system for closed-loop center pivot irrigation. ASABE Paper No. 002020. St. Joseph, MI: ASABE.
59. Wall RW, King BA (2004) Incorporating plug and play technology into measurement and control systems for irrigation management. ASABE Paper No. 042189. St. Joseph, MI: ASABE.
60. O'Shaughnessy SA, Evett SR (2010) Developing wireless sensor networks for monitoring crop canopy temperature using a moving sprinkler system as a platform. Appl Eng Agric 26(2):331–341
61. Vellidis G, Tucker M, Perry C, Kvien C, Bednarz C (2008) A real-time wireless smart sensor array for scheduling irrigation. Comput Electron Agric 61(1):44–50
62. Pierce FJ, Elliott TV (2008) Regional and on-farm wireless sensor networks for agricultural systems in eastern Washington. Comput Electron Agric 61(1):32–43
63. Diaz SE, Perez JC, Mateos AC, Marinescu MC, Guerra BB (2011) A novel methodology for the monitoring of the agricultural production process based on wireless sensor networks. Comput Electron Agric 76:252–265
64. Kim Y, Evans RG, Iversen WM (2008) Remote sensing and control of an irrigation system using a wireless sensor network. IEEE Trans Instrum Meas 57(7):1379–1387
65. Kim Y, Evans RG (2009) Software design for wireless sensor-based site-specific irrigation. Comput Electron Agric 66(2):159–165
66. Zhang Z (2004) Investigation of wireless sensor networks for precision agriculture. ASAE/CSAE annual international meeting. Paper No. 041154. St. Joseph, MI: ASAE
67. Oksanen T, Ohman M, Miettinen M, Visala A (2004) Open configurable control system for precision farming. ASABE Paper No. 701P1004. St. Joseph, MI: ASABE
68. Lee WS, Burks TF, Schueller JK (2002) Silage yield monitoring system. ASABE Paper No. 021165. St. Joseph, MI: ASABE
69. Dowla F (2006) Handbook of RF and wireless technologies. Elsevier Science, Burlington, MA

70. Li Y, Ephremides A (2007) A joint scheduling, power control, and routing algorithm for ad hoc wireless networks. Ad Hoc Netw 5(7):959–973
71. Demirkol I, Esroy C Energy and delay optimized contention for wireless sensor networks. Article in press as: Comput Netw
72. Hebel MA (2006) Meeting wide-area agricultural data acquisition and control challenges through Zigbee wireless network technology. In: Proceedings of international conference of computers in agriculture and natural resources. July 24-26, 2006. Lake Buena Vista, FL. 2006; pp 234–239
73. Goense D, Thelen J (2005) Wireless sensor networks for precise phytophthora decision support. In: Proceedings of ASAE annual international meeting, July 17-20, 2005, Tampa, Florida. Paper No. 053099. 2005
74. Andrade-Sanchez P, Pierce FJ, Elliott TV (2007) Performance assessment of wireless sensor networks in agricultural settings. St. Joseph, Mich.: ASABE, Paper No. 073076
75. Blonquist JM, Jones SB, Robinson DA (2006) Precise irrigation scheduling for turfgrass using a subsurface electromagnetic soil moisture sensor. Agric Water Manag 84:153–165
76. Huang YB, Thomson SJ, Lan YB, Maas SJ (2010) Multispectral imaging systems for airborne remote sensing to support agricultural production management. Int J Agric Biol Eng 3(1):50–62
77. Son NT, Chen CF, Chen CR, Chang L, Duc H, Nguyen L (2013) Prediction of rice crop yield using MODIS EVI-LAI data in the Mekong Delta, Vietnam. Int J Remote Sens 34(20):7275–7292
78. Johnson DM (2014) An assessment of pre- and within-season remotely sensed variables for forecasting corn and soybean yields in the United States. Remote Sens Environ 141(4):116–128
79. Bhattacharya BK, Chattopadhyay C (2013) A multi-stage tracking for mustard rot disease combining surface meteorology and satellite remote sensing. Comput Electron Agric 90:35–44
80. Jonas F, Gunter M (2007) Multi-temporal wheat disease detection by multi-spectral remote sensing. Precision Agric 8(3):161–172
81. Berk P, Hocevar M, Stajnko D, Belsak A (2016) Development of alternative plant protection product application techniques in orchards, based on measurement sensing systems: a review. Comput Electron Agric 124:273–288
82. Van HL, Tang X (2014) An efficient algorithm for scheduling sensor data collection through multi-path routing structures. J Netw Comput Appl 38(2):150–162
83. Hutchinson M, Oh H, Chen WH (2017) A review of source term estimation methods for atmospheric dispersion events using static or mobile sensors. Inf Fusion 36(11):130–148
84. Wu BF, Gommes R, Zhang M, Zeng HW, Yan NN, Zou WT et al (2015) Global crop monitoring: a satellite-based hierarchical approach. Remote Sens 7(4):3907–3933
85. Sherine M, Abd EK, Basma M, Mohammad EB (2013) Precision farming solution in Egypt using the wireless sensor network technology. Egypt Inform J 14:221–233
86. Aqeel-Ur-Rehman, Abbasi AZ, Islam N, Shaikh ZA (2014) A review of wireless sensors and networks applications in agriculture. Comput Stand Interfaces 36:263–270
87. Blonquist JM, Jones SB, Robinson DA (2006) Precise irrigation scheduling for turf grass using a subsurface electromagnetic soil moisture sensor. Agric Water Manag 84:153–165
88. Dias PC, Roque W, Ferreira EC, Siqueira Dias JA (2013) A high sensitivity single-probe heat pulse soil moisture sensor based on a single non junction transistor. Comput Electron Agric 96:139–147
89. Xiao D, Feng J, Wang N, Luo X, Hu Y (2013) Integrated soil moisture and water depth sensor for paddy fields. Comput Electron Agric 98:214–221
90. Kim Y, Evans RG, Iversen WM (2009) Evaluation of closed-loop site-specific irrigation with wireless sensor network. J Irrig Drain Eng ASCE 135(1):25–31
91. Cao H, Guo FT, Fan YS, Duan FY, Han QB, Jia YH et al (2016) Running speed and pressure head loss of the light and small sprinkler irrigation system. J Drain Irrig Mach Eng 34(2):179–184 (in Chinese)

92. Cai SB, Zhu DL, Ge MS, Liu KN, Li D (2017) Photovoltaic optimization of solar-powered linear move sprinkler irrigation system. J Drain Irrig Mach Eng 35(5):417–423 (in Chinese)
93. Bautista-Capetillo C, Robles O, Salinas H, Playán E (2014)A particle tracking velocimetry technique for drop characterization in agricultural sprinklers. Irrig Sci 32(6): 437–447
94. Sayyadi H, Nazemi AH, Sadraddini AA, Delirhasannia R (2014) Characterising droplets and precipitation profiles of a fixed spray-plate sprinkler. Biosyst Eng 119(1):13–24
95. Liu JP, Liu XF, Zhu XY, Yuan SQ (2016) Droplet characterisation of a complete fluidic sprinkler with different nozzle dimensions. Biosyst Eng 148(6):90–100
96. Liu JP, Yuan SQ, Darko RO (2016) Characteristics of water and droplet size distributions from fluidic sprinklers. Irrig Drain 65(4):522–529
97. Zhu XY, Yuan SQ, Jiang JY, Liu JP, Liu XF (2015) Comparison of fluidic and impact sprinklers based on hydraulic performance. Irrig Sci 33(5):367–374
98. Zhang L, Merley GP, Pinthong K (2013) Assessing whole-field sprinkler application uniformity. Irrig Sci 31:87–105
99. Dwomoh FA, Yuan S, Hong L (2013) Field performance characteristics of fluidic sprinkler. Appl Eng Agric 29(4):529–536
100. Karmeli D, Peri G (1974) Basic principles of pulse irrigation. J Irrig Drain Div ASCE 100(IR3):309–319
101. Wang YX, Xu SS, Li WB, Kang F, Zheng YJ (2017) Identification and location of grapevine sucker based on information fusion of 2D laser scanner and machine vision. Int J Agric Biol Eng 10(2):84–93
102. Evans RG, Harting GB (1999) Precision irrigation with center pivot systems on potatoes. In: Walton R, Nece RE (eds) Proceedings of ASCE 1999 international water resources engineering conference, Reston, VA: ASCE, 1999; CD-ROM
103. Evans RG, Han S, Schneider SM, Kroeger MW (1996) Precision center pivot irrigation for efficient use of water and nitrogen. In: Roberts PC, Rust RH, Larsen WE. Madison WI (eds) Proceedings of the 3rd international conference on precision agriculture. ASA-CSSA, pp 75–84
104. Bao Y, Liu JP, Liu XF, Tian K, Zhang Q (2016) Experimental study on effects of pressure on water distribution model of low-pressure sprinkler. J Drain Irrig Mach Eng 34(1):81–85 (in Chinese)
105. Chávez JL, Pierce FJ, Evans RG (2010) Compensating inherent linear move water application errors using a variable rate irrigation system. Irrig Sci 28(3):203–210
106. Zhu X, Chikangaise P, Shi W, Chen W, Yuan S (2018) Review of intelligent sprinkler irrigation technologies for remote autonomous system. Int J Agric Biol Eng 11:23–30

Printed in the United States
by Baker & Taylor Publisher Services